# Les Knight
## Australia's Dambuster
## Marcus Fielding

ECHO BOOKS

First published in 2019 by Barrallier Books Pty Ltd,
trading as Echo Books
Registered Office: 35—37 Gordon Avenue, West Geelong, Victoria 3220, Australia.
www.echobooks.com.au
Copyright ©Marcus Fielding
Creator: Fielding, Marcus.
Title: /Les Knight: Australia's Dambuster/MarcusFielding.
ISBN: 9780648355281 (hardback)

A catalogue record for this book is available from the National Library of Australia

Book layout and design by Peter Gamble, Canberra.
Set in Garamond Premier Pro Display, 12/17 and/Trajan Pro.
www.echobooks.com.au
Cover image: *Canal Sortie (armed with Upkeep)*, Gary Easton
https://www.flightartworks.com

# Contents

| | |
|---|---|
| Introduction | v |
| An Unlikely Warrior | 1 |
|    No. 50 Squadron RAF | 6 |
|    Les Knight's Crew Mates | 10 |
|    The Battle of the Ruhr | 13 |
|    Sir Arthur 'Bomber' Harris | 16 |
|    Avro Lancaster | 20 |
|    Barnes Wallis | 22 |
|    Guy Gibson—Part I | 26 |
| 617 Squadron | 43 |
|    No. 617 Squadron RAF | 46 |
|    RAF Scampton | 49 |
|    The Bouncing Bomb | 52 |
| The Dams Raid Briefing | 63 |
|    The Möhne Dam | 68 |
|    The Eder Dam | 71 |
|    Routes | 73 |
|    Codewords | 74 |
| The Dams Raid | 75 |
| The Möhne Dam Attack | 81 |
| The Eder Dam Attack | 85 |
| The Sorpe Dam Attack | 91 |
| Returning to RAF Scampton | 95 |

| | |
|---|---|
| Accolades | 97 |
| Impact | 103 |
|     Guy Gibson—Part II | 109 |
| The Dortmund–Ems Canal Attack | 121 |
| Les' Legacy | 129 |
| Remembering Les Knight | 133 |
| Conclusion | 139 |
| About the Author | 142 |

# Introduction

In 2019 there is precious little living memory of the Second World War. Very few veterans remain and their memories are failing. Seventy-four years after the conclusion of the war any appreciation of the demands and atmosphere of those days is largely limited to historians who can only try to comprehend the scale and impact of events with the benefit of reading and study. Attempting to relay that knowledge to a young man in his early twenties today is very difficult. That is why I believe this book is important. It tells the story of a seemingly ordinary young man who in the space of just two short years was motivated and capable of extraordinary deeds—such were the times.

Les Knight became a 'household name' in Australia in 1943. Amongst the thousands of young men who had volunteered and gone to war this particular young man, seemingly out of nowhere, leapt into the public consciousness as a result of the Dams Raid. And as a result, he was not only awarded a decoration for distinguished service in combat but he met the King and Queen in the process. His skill and courage were admired by many; and he was just 22 years old.

This is the story of Les Knight's rapid and remarkable journey to fame. But it is also the story of his equally tragic death just a few months later.

Such was the randomness of war; earlier achievements counted for little when the immense risks of flying a bombing mission over enemy territory were concerned. But even in his final actions on earth Les demonstrated immense bravery and selflessness.

Australians would benefit to know a little more about the people that fought in the Second World War. Les Knight is a tremendous example of what people are capable of achieving in extraordinary circumstances. His memory is held in high regard in a small village in Holland and we in Australia should venerate him as well.

# An Unlikely Warrior

Leslie Gordon Knight was born on 7 March 1921. He was the older of the two sons of William Henry Harold 'Harry' Knight and Nellie Marsom Knight. They lived at 51 Bowen Street in Camberwell in Melbourne, Victoria. His father, an engineer by trade, later said that Les had no aptitude for tools and no mechanical skills. Les was not active in sports but he was studious in his studies and matriculated with his intermediate and leaving certificate. Les had intended to become an accountant, and worked as a clerk in the accounting firm of Davey Balding and Co., which had offices at 31 Queen Street in Melbourne city.

Les' Air Force enlistment file portrait photo.

On Monday 3 February 1941—just one month short of his twentieth birthday—Les walked down Queen Street and walked into the door of No. 1 Recruiting Centre at 108 Queens Street. There he enlisted as trainee aircrew for the duration of the War plus 12 months. He was assigned the service number 401499.

The preceding Thursday Adolf Hitler had given a speech before 18,000 people at the Berlin Sportpalast on the eighth anniversary of the Nazis' coming to power. Hitler declared that any ship carrying aid to England within the range of German U—boats would be torpedoed, and also warned the United States that if anyone on the American continent tried to interfere in the European conflict, Germany's war aims would quickly change. The weekend papers would no doubt have carried advice of this speech and it might have had a bearing on Les' decision to sign up.

*Les' formal portrait photo at his initial training course.*

*Les displaying his pilots' wings after graduating in August 1941.*

Les was immediately shipped out to No. 1 Initial Training School (ITS) at Somers, Victoria and then graduated on 26 April 1941 as a Leading Aircraftman. At this stage he was identified as being suitable for training as a pilot—probably because he had matriculated.

On 1 May 1941 he was posted to the No. 7 Elementary Flying Training School (EFTS) at the Western Junction Aerodrome in Tasmania where he learned to fly in the de Havilland DH.82 Tiger Moth single engined biplane. On 31 May 1941, when Les was training at the school two No. 7 EFTS Tiger Moths collided as they were taking off. The pilot of one of the two aircraft was killed, and his passenger was injured. This training accident would have had a significant impact on all the students and staff at the school.

Graduating from No. 7 EFTS Les was posted to the No. 1 Service Flying Training School (SFTS) at Point Cook, in Victoria on 30 June 1941. He passed the Intermediate Flying Course (#12) on 21 August 1941 and was awarded the Flying Badge. He then passed the Advanced Flying Course on 16 October 1941 and promoted to Temporary Sergeant. At Point Cook Les learned to fly twin engined Avro Ansons.

Les was then assigned to the No. 1 Embarkation Depot on 17 October 1941 in anticipation of serving in the United Kingdom. He was attached to the Royal Air Force on 12 November 1941 and embarked on a troop ship at Newcastle on 17 November 1941.

Les and his mother Nellie before his embarkation.

Disembarking on 4 February 1942 Les marched in to No. 3 Personnel Reception Centre on 5 February 1942. From there he was assigned to the No. 15 Advanced Flying Unit at RAF Cottesmore on 3 March 1942—just in time to celebrate his twenty—first birthday.

After a month at RAF Cottesmore he moved north to RAF Kirmington where he gained hours flying Airspeed AS.10 Oxfords—a twin engined aircraft that was used to train crews. On 17 April 1942 Les was promoted to Temporary Flight Sergeant—most probably after passing another training requirement. At this stage Les would have been introduced to airmen from all over the Empire—Brits, Scots, Welsh, Canadians and New Zealanders. Trainees would have been moved continually between crews in order for them to get used to working with different people and to see how well they worked in teams.

On 28 May 1942 Les was assigned to the No. 14 Operational Training Unit at RAF Cottesmore. This unit was part of RAF Bomber Command and tasked with training night bomber crews in the Handley Page HP.52 Hampden—a twin—engined medium bomber.

Les had a family contact in the UK. During the First World War Les' father while making munitions in Barrow met Newton. After Les' parents emigrated to Australia they had remained in communication. The two

*Group photo of one of Les' training courses.*

families agreed that his young son could treat their home as his base in England while on leave. Les visited the family in Ambleside in Cumbria whenever he had a few days of leave. Vera Newton was just 12 but remembers him fondly: 'He never drank, smoked or swore.'

*Vera Newton with her 'older brother' Les.*

On 14 August 1942 Les was posted to the No. 1654 Conversion Unit at RAF Wigsley in Nottinghamshire where he was qualified to fly Avro Lancasters. The Lancaster, was basically an Avro Manchester fuselage mounted to a revised main wing that held four Rolls—Royce Merlin engines rather than two Vulture engines. The redesign allowed Avro to produce an aircraft that had good handling characteristics, range, immense bomb lifting capability and most importantly was liked by the aircrews who flew them. It was this training period that a crew formed around Les as the pilot.

Finally, on 18 September 1942 after 20 months of training, Les and his crew were posted to an operational unit—No. 50 Squadron at RAF Swinderby. From being a raw recruit in Melbourne with no particular mechanical aptitude he was now the pilot of the world's most technologically advanced heavy bomber and about to take part in the air war against Nazi Germany; and he was just 21 years old.

Since Les enlisted the Nazis had invaded Russia and seemed unstoppable. The Japanese had attacked Pearl Harbour and captured key parts of the western Pacific. Australia felt very much threatened. That week the British war film 'In Which We Serve', directed by Noël Coward, was released in the United Kingdom and we can only imagine that Les went to see this and re-doubled his commitment to serving with vigour and honour.

On 1 October 1942 Les flew his first operational mission to Wismar on the Baltic coast. He quickly got used to the Lancaster and the routine of flying operational sorties which included bombing and laying sea mines—mostly at night. Laying sea mines was colloquially referred to as 'gardening'.

One of the reasons the crew worked so well together might have been its nationally diverse nature. There were two slightly older Englishmen, Sidney Hobday and Ed Johnson. The Flight Engineer Ray Grayston was also English but Bob Kellow, the wireless operator, was Australian and both the gunners, Fred Sutherland and Harry O'Brien were Canadians. All of them shared the highest regard for their young pilot.

In October 1942 50 Squadron returned to RAF Skellingthorpe near Lincoln which they had vacated four months earlier to allow for the runways to be extended and the base made capable for heavy bombers.

On 8 December 1942 Les was commissioned as a Pilot Officer and by March 1943 Les and his crew had flown on some twenty—six operations.

## NO. 50 SQUADRON RAF

*The 50 Squadron RAF Crest.*

No. 50 Squadron of the Royal Flying Corps founded at Dover on 15 May 1916. It was equipped with a mixture of aircraft, including Royal Aircraft Factory B.E.2s and Royal Aircraft Factory B.E.12s in the home defence role, having flights based at various airfields around Kent. It flew its first combat mission in August 1916, when its aircraft helped to repel a German Zeppelin. On 7 July 1917 a 50 Squadron Armstrong Whitworth F.K.8 shot down a German Gotha bomber off the North Foreland of Kent. In February 1918, it discarded its miscellany of aircraft to standardise on the more capable Sopwith Camel fighter, continuing to defend Kent.

By October 1918, it was operating its Camels as night fighters. It was during this period that the squadron started using the 'running dogs' device on squadron aircraft, a tradition that continued until 1984. The device arose from the radio call sign Dingo that the squadron was allocated as part of the Home Defence network. It disbanded on 13 June 1919. The last Commanding Officer of the squadron before it disbanded was Major Arthur Harris later to become Air Officer Commanding-in-Chief of RAF Bomber Command during the Second World War.

No. 50 Squadron reformed at RAF Waddington on 3 May 1937, equipped with Hawker Hind biplane light bombers. It started to convert to the Handley Page Hampden monoplane medium bomber in December 1938, discarding its last Hinds in January 1939. It was still equipped with Hampdens when the Second World War broke out, forming part of 5 Group, Bomber Command. It flew its first bombing raid on 19 March 1940 against the seaplane base at Hörnum on the island of Sylt.

On 12 April 1940, in attempt to attack German warships off Kristiansand returning from the German invasion of Norway, 50 Squadron took part in what was the largest British air raid of the war so far, with a total of 83 RAF bombers attempting to attack the German fleet. When 12 Hampdens of 50 and 44 Squadron spotted a German warship and attempted to attack, they lost 6 of their number to beam attacks by German fighters, with 13 officers and men from 50 Squadron dead or missing. After these losses, daylight attacks with Hampdens were abandoned.

50 Squadron continued operations by night, taking part in the RAF's strategic bombing offensive against the Germans through the remainder of 1940 and 1941. It re-equipped with Avro Manchesters from April 1942. The Manchester was disappointing, however, with unreliable engines and had a lower ceiling than the Hampden it replaced. Despite these problems, 50 Squadron continued in operations, contributing 17 Manchesters to 'Operation Milliennium' the '1,000 aircraft' raid against Cologne on 30/31 May 1942. It lost two aircraft that night, one of which piloted by Flying Officer Leslie Thomas Manser who was posthumously awarded the Victoria Cross for pressing on with the attack

after his aircraft was heavily damaged, and when a crash became inevitable, sacrificing his own life by remaining at the controls to allow the rest of his crew to parachute to safety.

The Squadron soon re-equipped with the four-engined Avro Lancaster, which it used for the rest of the war against German targets, flying its last mission of the war against an oil refinery at Vallø in Norway on 25/26 April 1945. The squadron flew 7,135 sorties during the war with a loss of 176 aircraft. It replaced its Lancasters with Avro Lincolns in 1946, disbanding at Waddington on 31 January 1951.

50 Sqn and the Handley Page HP.52 Hampden circa 1940.

The Handley Page HP.52 Hampden.

*50 Sqn at RAF Skellingthorpe.*

*The Avro Manchester Mk I.*

## Les Knight's Crew Mates

### Flying Officer Harold Hobday (RAF) — Navigator

Harold Sydney Hobday was born in Croydon, Surrey, UK on 28 January 1912. After leaving school he worked in the aviation department of Lloyd's the insurance business. After joining the ARF in 1940 he underwent part of his training in South Africa before qualifying as a navigator in early 1942 and then being commissioned. In the summer of 1942 he crewed up during training with Les Knight and they joined 50 Squadron in September 1942.

Although some eight years older than his young Australian skipper (still then a sergeant pilot) they obviously boded well and flew on some 25 operations together up until March 1943, when the whole crew volunteered to be transferred to the new squadron at Scampton.

### Flying Officer Edward Johnson (RAF) — Bomb Aimer

Edward Cuthbert Johnson was born in Lincoln, UK on 3 May 1912. When his father was killed on the Western Front his mother moved to Gainsborough, although he was educated at Lincoln Grammar School. On leaving school Johnson worked for Woolworths and then the catering firm Lyons, and then in a boarding house business in Blackpool with his wife's family.

Johnson joined the RAF in 1940, qualified as an observer/bomb aimer in early 1942, and was commissioned. After further training he was posted briefly to 106 Squadron, but then sent back to a training unit to be crewed up with Les Knight and his colleagues. They moved to 50 Squadron in September 1942, and Johnson flew on some 22 operations with the Les Knight crew.

Johnson and Hobday were the elder statemen of the Knight crew, both nine years older than their skipper, and senior to him in rank. But they worked well as a team, each obviously seeing in the younger man the qualities of an outstanding pilot.

Sergeant Raymond Grayston (RAF) — Flight Engineer

Raymond Ernest Grayston was born on 13 October 1918 at Dunsfold, Surry, UK and worked as a automobile engineers before the war. Like many young men of his generation Grayston was fascinated by flying and volunteered for the RAF at the start of the war. He later described how much he enjoyed riding a motorbike at speed and this was one of the things that attracted him to the air force. Initially he served as ground crew but then, along with others who were mechanically minded, he was selected to train as a flight engineer on the new generation of heavy bombers which had larger crews. He was posted to 50 Squadron in October 1942 and teamed up with Les Knight, who had started a tour of operations about a month earlier with a different flight engineer. Grayston flew on some eighteen operations with Les before the whole crew volunteered to be transferred to the new squadron at Scampton.

Flight Sergeant Robert Kellow (RAAF) — Wireless Operator

Robert George Thomas Kellow was born in Newcastle, New South Wales, Australia on 13 December 1916. He went to Newcastle High School and worked as a shop assistant after leaving school. He enlisted in the RAAF and was selected to train as a wireless operator/air gunner and was sent to Canada for training. He was then posted to the UK and arrived in January 1942.

After further training Kellow crewed up with Les Knight and then posted to 50 Squadron. He completed some 25 operations with Les before the whole crew volunteered to be transferred to the new squadron at Scampton. Shortly before the transfer came about Kellow was recommended for a Distinguished Flying Cross and commissioned.

**Flight Sergeant Frederick Sutherland (RCAF) — Air Gunner**

Frederick Edwin Sutherland was born in Peace River, Alberta, Canada on 26 February 1923. From a young age he wanted to fly and had dreams of becoming a bush pilot. He joined the RCAF in 1941 as soon as he turned 18. After initial training he volunteered for air gunner duties.

Sutherland arrived in England in 1942 and crewed up with Les Knight and his colleagues at a training unit before they were all posted to 50 Squadron in September 1942. Sutherland flew 25 operations with Knight before the whole crew volunteered to transfer to the new squadron in March 1943.

**Flight Sergeant Henry O'Brien (RCAF) — Air Gunner**

Henry Earl O'Brien was born in Regina, Saskatchewan, Canada on 15 August 1922. He volunteered for the RCAF as soon as he turned 18 and was selected for gunnery training.

On arriving in the UK and further training he crewed up with Les Knight and his colleagues. They were all posted to 50 Squadron in September 1942. O'Brien flew 23 operations with Les and the crew.

## The Battle of the Ruhr

The Battle of the Ruhr of 1943 was a five-month British campaign of strategic bombing during the Second World War against the Nazi Germany Ruhr Area, which had coke plants, steelworks, and 10 synthetic oil plants.

The campaign bombed 26 major Combined Bomber Offensive targets. The targets included the Krupp armament works (Essen), the Nordstern synthetic-oil plant, and the Rheinmetall–Borsig plant in Düsseldorf. The latter was safely evacuated during the Battle of the Ruhr.

Although not strictly part of the Ruhr area, the battle of the Ruhr included other cities such as Cologne which were within the Rhine-Ruhr region and considered part of the same 'industrial complex'. Some targets were not sites of heavy industrial production but part of the production and movement of materiel.

Although the Ruhr had always been a target for the RAF from the start of the war, the organized defences and the large amount of industrial pollutants produced that gave a semi-permanent smog or industrial haze hampered accurate bombing. Before the Battle of the Ruhr ended, Operation Gomorrah began the 'Battle of Hamburg'. Even after this switch of focus to Hamburg, there would be further raids on the Ruhr area by the RAF—in part to keep German defences dispersed, just as there had been raids on areas other than the Ruhr during the battle.

The British bomber force consisted mainly of the twin-engined Vickers Wellington medium bomber and the four-engined 'heavies', the Short Stirling, Handley Page Halifax and Avro Lancaster. The Wellington and Stirling were the two oldest designs and limited in the type or weight of bombs carried. The Stirling was also limited to a lower operational height. Bombers could carry a range of bombs—Medium Capacity bombs of about 50% explosive by weight, High Capacity 'Blockbusters' that were mostly explosive, and incendiary devices. The combined use of the latter two were most effective in setting fires in urban areas.

British raids were by night—the losses in daylight raids having been too heavy to bear. By this point in the war, RAF Bomber Command were using navigation aids, the Pathfinder force and the bomber stream tactic together. Electronic navigation aids such as 'Oboe', which had been tested against Essen in January 1943, meant the Pathfinders could mark the targets despite the industrial haze and cloud cover that obscured the area by night. Guidance markers put the main force over the target area, where they would then drop their bombloads on target markers.

The bomber stream concentrated the force of bombers into a small-time window, such that it overwhelmed fighter defences in the air and firefighting attempts on the ground. For most of the Battle of the Ruhr the Oboe de Havilland Mosquitoes came from one squadron, No. 109. The number of Oboe aircraft that could be used at any time was limited by the number of ground stations.

The US Army Air Force (USAAF) had two 4-engined heavy bombers available: the Boeing B-17 Flying Fortress and Consolidated B-24 Liberator—neither of these American heavy bomber designs had a bomb bay suitable to carry the RAF's blockbuster bombs or anything comparable. USAAF raids were by daylight, the closely massed groups of bombers covering each other with defensive fire against fighters. Between them, the Allies could mount 'round the clock' bombing. The USAAF forces in the UK were still increasing during 1943 and the majority of the bombing was by the RAF.

The German defence consisted of anti-aircraft weapons and day and night fighters. The Kammhuber Line used radar to identify the bomber raids and then controllers directed night fighters onto the raiders. During the battle of the Ruhr, Bomber Command estimated about 70% of their aircraft losses were due to fighters. By July 1943, the German night fighter force totalled 550.

Through the summer of 1943, the Germans increased the ground-based anti-aircraft defences in the Ruhr Area; by July 1943 there were more than 1,000 large flak guns (88 mm calibre guns or greater) and 1,500 lighter guns (chiefly 20 mm and 37 mm calibre). This was about one-third of all anti-aircraft guns in Germany. Six-hundred thousand personnel were required to man the AA defences of Germany. The British crews called the area sarcastically 'Happy Valley' or the 'Valley of no Return'.

In his study of the German war economy, Adam Tooze stated that during the Battle of the Ruhr, Bomber Command severely disrupted German production. Steel production fell by 200,000 tons. The armaments industry was facing a steel shortfall of 400,000 tons. After doubling production in 1942, production of steel increased only by 20 percent in 1943. Hitler and Speer were forced to cut planned increases in production.

This disruption resulted in the 'Zulieferungskrise' (sub-components crisis). The increase of aircraft production for the Luftwaffe also came to an abrupt halt. Monthly production failed to increase between July 1943 and March 1944. 'Bomber Command had stopped Speer's armaments miracle in its tracks'. At Essen after more than 3,000 sorties and the loss of 138 aircraft, the 'Krupp works… and the town…itself contained large areas of devastation', and Krupp never restarted locomotive production after the second March raid.

## Sir Arthur 'Bomber' Harris

Arthur Harris was born in Cheltenham, Gloucestershire, England on April 13th 1892. Harris emigrated to Southern Rhodesia in 1910, aged 17, but returned to England in 1915 to fight in the European theatre of the First World War. He joined the Royal Flying Corps, with which he remained until the formation of the Royal Air Force in 1918. He served with territorial defence against Zeppelin attacks. He later fought against German bombers in Britain. By the end of the First World War, he had been squadron leader for some time and had been awarded the Air Force Cross.

Harris remained in the Air Force through the 1920s and 1930s, serving in India, Mesopotamia, Persia, Egypt, Palestine, and elsewhere. In September 1939, the RAF appointed Harris to lead Bomber Commands 5 Group. Later in November 1940 he left operational command to join the Air Ministry as Deputy Chief of Air Staff. On 23rd February 1942 Harris was appointed Commander-in-Chief, Bomber Command; a post he held until 1945.

The Butt Report, circulated in August 1941, found that in 1940 and 1941 only one in three attacking aircraft got within five miles (eight kilometres) of their target. As part of the response Harris was appointed Commander-in-Chief (C-in-C) of Bomber Command in February 1942.

In 1942, Professor Frederick Lindemann (later ennobled as Lord Cherwell), having been appointed the British government's leading scientific adviser (with a seat in the Cabinet) by his friend, Prime Minister Winston Churchill, presented a seminal paper to Cabinet advocating the area bombing of German cities in a strategic bombing campaign. It was accepted by Cabinet and Harris was directed to carry out the task (area bombing directive). It became an important part of the total war waged against Germany.

At the start of the bombing campaign, Harris said, quoting the Old Testament: 'The Nazis entered this war under the rather childish delusion that they were going to bomb everyone else, and nobody was going to bomb them. At Rotterdam, London, Warsaw and half a hundred other places, they put their rather naive theory into operation. They sowed the wind, and now they are going to reap the whirlwind.'

At first the effects were limited because of the small numbers of aircraft used and the lack of navigational aids, resulting in scattered, inaccurate bombing. As production of better aircraft and electronic aids increased, Harris pressed for raids on a much larger scale, each to use 1,000 aeroplanes. In Operation Millennium Harris launched the first RAF 'thousand bomber raid' against Cologne (Köln) on the night of 30/31 May 1942. This operation included the first use of a bomber stream, which was a tactical innovation designed to overwhelm the German night-fighters of the Kammhuber Line.

Harris was just one of an influential group of high-ranking Allied air commanders who continued to believe that massive and sustained area bombing alone would force Germany to surrender. On a number of occasions he wrote to his superiors claiming the war would be over in a matter of months, first in August 1943 following the tremendous success of the Battle of Hamburg (codenamed Operation Gomorrah), when he assured the Chief of the Air Staff, Sir Charles Portal, that his force would be able 'to produce in Germany by April 1st 1944 a state of devastation in which surrender is inevitable', and then again in January 1944. Winston Churchill continued to regard the area bombing strategy with distaste, and official public statements still maintained that Bomber Command was attacking only specific industrial and economic targets, with any civilian casualties or property damage being unintentional but unavoidable.

In October 1943, emboldened by his success in Hamburg and increasingly irritated with Churchill's hesitance to endorse his tactics wholeheartedly, Harris urged the government to be honest with the public regarding the purpose of the bombing campaign: 'The aim of the Combined Bomber Offensive...should be unambiguously stated [as] the destruction of German cities, the killing of German workers, and the disruption of civilised life throughout Germany...the destruction of houses, public utilities, transport and lives, the creation of a refugee problem on an unprecedented scale, and the breakdown of morale both at home and at the battle fronts by fear of extended and intensified bombing, are accepted and intended aims of our bombing policy. They are not by-products of attempts to hit factories. However, at this time many senior Allied air commanders still thought area bombing was less effective.

In November 1943 Bomber Command began what became known as the Battle of Berlin: a series of massive raids on Berlin that lasted until March 1944. Harris sought to duplicate the victory at Hamburg, but Berlin proved to be a far more difficult target. Although severe general damage was inflicted, the city was much better prepared than Hamburg, and no firestorm was ever ignited. Anti-aircraft defences were also extremely effective and bomber losses were high; during this time the British lost 1,047 bombers, with a further 1,682 damaged, culminating in the disastrous raid on Nuremberg on 30 March 1944, when 94 bombers were shot down and 71 damaged, out of 795 aircraft.

Harris's continued preference for area bombing over precision targeting remains controversial, partly because many senior Allied air commanders thought it less effective and partly for the large number of civilian casualties and destruction this strategy caused in Continental Europe. After the war Harris moved to South Africa where he managed the South African Marine Corporation.

Whenever the bombing campaign of the Second World War is considered it must be appreciated that the war was an 'integrated process'. As an example, quoting Albert Speer from his book Inside the Third Reich, 'ten thousand [88mm] anti-aircraft guns…could well have been employed in Russia against tanks and other ground targets.'

At the end of the war Harris was promoted to RAF Marshal. He retired from the RAF in 1946 to pursue a successful business career in South Africa. He later returned to Britain and was made a baronet. In 1992, eight years after his death, a monument to him was erected in central London.

*Sir Arthur 'Bomber' Harris.*

## Avro Lancaster

The Avro Lancaster was a British four-engined Second World War heavy bomber. It was designed and manufactured by Avro as a contemporary of the Handley Page Halifax, both bombers having been developed to the same specification, as well as the Short Stirling, all three aircraft being four-engined heavy bombers adopted by the Royal Air Force (RAF) during the same wartime era.

The Lancaster has its origins in the twin-engine Avro Manchester which had been developed during the late 1930s in response to the Air Ministry Specification P.13/36 for a capable medium bomber for 'world-wide use'. Originally developed as an evolution of the Manchester (which had proved troublesome in service and was retired in 1942), the Lancaster was designed by Roy Chadwick and powered by four Rolls-Royce Merlins and in one version, Bristol Hercules engines.

*The Avro Lancaster Mk I.*

The Lancaster first saw service with RAF Bomber Command in 1942 and as the strategic bombing offensive over Europe gathered momentum, it was the main aircraft for the night-time bombing campaigns that followed. As increasing numbers of the type were produced, it became the principal heavy bomber used by the RAF, the RCAF and squadrons from other Commonwealth and European countries serving within the RAF, overshadowing contemporaries such as the Halifax and Stirling.

A long, unobstructed bomb bay meant that the Lancaster could take the largest bombs used by the RAF, including the 4,000 lb (1,800 kg), 8,000 lb (3,600 kg) and 12,000 lb (5,400 kg) Blockbusters, loads often supplemented with smaller bombs or incendiaries. The 'Lanc', as it was affectionately known, became one of the more famous and most successful of the Second World War night bombers, delivering 608,612 long tons of bombs in 156,000 sorties.

The versatility of the Lancaster was such that it was chosen to equip 617 Squadron for 'Operation Chastise' - the attack on German Ruhr valley dams. The aircraft was modified to carry the 'bouncing bomb' designed by Barnes Wallis. Known as the Type 464 (Provisioning) or later the B.III (Special), twenty-three aircraft had their bomb bay doors were removed and the ends of the bay were covered with fairings. The weapon was suspended on pivoted, vee-shaped struts which sprang apart when the bomb-release button was pressed. A drive belt and pulley to rotate the bomb at 500 rpm was mounted on the starboard strut and driven by a hydraulic motor housed in the forward fairing. The mid-upper turret was removed to save weight and the gunner moved to the front turret to relieve the bomb aimer from having to man the front guns so that he could assist with map reading. A more bulbous bomb aimer's blister was fitted and lamps were fitted in the bomb bay and nose for the simple height measurement system which enabled the accurate control of low-flying altitude at night. Vickers modified the aircraft at Woodford Aerodrome near Stockport where the workers worked day and night. The first adapted aircraft arrived at Scampton on 8 April, 1943.

*The Avro Lancaster fitted with a Bouncing Bomb.*

A practice 'Upkeep' bouncing bomb mounted under Wing Commander Guy Gibson's Avro Type 464 (Provisioning) Lancaster, ED932/G 'AJ-G', at Manston, Kent, while conducting dropping trials off Reculver. The chain was driven by a hydraulic motor and gave the bomb its backspin.

Although the Lancaster was primarily a night bomber, it excelled in many other roles, including daylight precision bombing, for which some Lancasters were adapted to carry the 12,000 lb (5,400 kg) 'Tallboy' and then the 22,000 lb (10,000 kg) 'Grand Slam' earthquake bombs (also designed by Wallis). This was the largest payload of any bomber in the war.

## BARNES WALLIS

Barnes Wallis was born in Ripley, Derbyshire and was educated at Christ's Hospital in Horsham and Haberdashers' Aske's Hatcham Boys' Grammar School in southeast London. He left school in 1904 with no qualifications but went to train as a marine engineer at an engineering company on the Isle of Wight. After several periods of unemployment, Wallis went to work for Vickers as Chief Assistant of Airship Design in 1913.

The R.9, R.23 and R.26 airships were all in service by the end of World War I in which he briefly served as a private. He designed the R.80 airship in 1924, this was said to be the best of its day. Wallis later designed the R.100 which used his revolutionary geodetic design principle. This saved weight and allowed the R.100 to be larger than any other airship. In 1930, the R.100 made a successful maiden voyage to Canada.

The R.100 built by Vickers was a technical success and may have revolutionised air transport. Unfortunately, the Government produced a rival airship to the R.100 which was ill designed and constructed too quickly in an attempt to catch up. Too many corners were cut and tragically on its maiden voyage the airship crashed in France and killed 48 crew. This incident represented a huge loss in national pride and confidence in airships. The R.100 was tarred with the same brush and the entire airship industry was scrapped in favour of the aeroplane industry which had made massive advancements.

In 1922 Wallis took a degree in engineering via the University of London External Programme. He continued work at Vickers but moved to aeroplane design becoming Chief Structures Designer. Wallis used his geodetic design principle from airships in aeroplane design. This type of construction could withstand greater stresses than conventional airframe designs and also greater damage as later air combat proved. Although the geodetic design seemed more complex, semi-skilled workers in wartime were able to produce the lattice airframes at an incredible rate.

The first aeroplane Wallis designed which used the geodetic construction was the Wellesley in 1935. This was later redesigned and became the famous Wellington bomber in 1936 as the prospect of war loomed. Use of the geodetic construction was met with fierce opposition by many high powers who favoured a more conventional design approach. Both the Wellesley and Wellington proved the advantage of the geodetic construction. The Wellington bomber was able to carry double the bomb load twice the distance than was agreed on the original contract specification. Vickers manufactured 11,460 Wellington bombers - more than any other British bomber in history.

After the outbreak of the Second World War in Europe in 1939, Wallis saw a need for strategic bombing to destroy the enemy's ability to wage war and he wrote a paper entitled 'A Note on a Method of Attacking the Axis Powers'. Referring to the enemy's power supplies, he wrote: "If their destruction or paralysis can be accomplished, they offer a means of rendering the enemy utterly incapable of continuing to prosecute the war." As a means to do this, he proposed huge bombs that could concentrate their force and destroy targets which were otherwise unlikely to be affected. Wallis's first super-large bomb design came out at some ten tonnes, far more than any current bomber could carry. Rather than drop the idea, this led him to suggest a plane that could carry it — the 'Victory Bomber'.

Early in 1942, Wallis began experimenting with skipping marbles over water tanks in his garden, leading to his April 1942 paper 'Spherical Bomb — Surface Torpedo'. The idea was that a bomb could skip over the water surface, avoiding torpedo nets, and sink directly next to a battleship or dam wall as a depth charge, with the surrounding water concentrating the force of the explosion on the target. A crucial innovation was the addition of backspin, which caused the bomb to trail behind the dropping aircraft (decreasing the chance of that aircraft being damaged by the force of the explosion below), increased the range of the bomb, and also prevented it from moving away from the target wall as it sank.

After some initial scepticism, the Air Force accepted Wallis's bouncing bomb (codenamed Upkeep) for attacks on the Möhne, Eder and Sorpe dams in the Ruhr area. The Möhne and Eder dams were successfully breached in a raid in May 1943 ('Operation Chastise') was, causing damage to German factories and disrupting hydro-electric power.

After the success of the bouncing bomb, Wallis was able to return to his huge bombs, producing first the Tallboy (6 tonnes) and then the Grand Slam (10 tonnes) deep-penetration earthquake bombs. These were not the same as the 5-tonne 'blockbuster' bomb, which was a conventional blast bomb. Although there was still no aircraft capable of lifting these two bombs to their optimal release altitude, they could still be dropped from a lower height, entering the earth at supersonic speed and penetrating to a depth of 20 metres before exploding. They were used on strategic German targets such as V-2 rocket launch sites, the V-3 supergun bunker, submarine pens, and other reinforced structures, large civil constructions such as viaducts and bridges, as well as the German battleship *Tirpitz*. They were the forerunners of modern bunker-busting bombs.

After the war he led aeronautical research and development at the British Aircraft Corporation until 1971. Unfortunately, due to financial limitations and political 'red tape' his obvious genius on later projects was never allowed to flourish. He died on 30th October 1979. Sir Barnes Neville Wallis was one of Britain's best engineers and inventors. An incredible man who certainly did much to help shorten and win the Second World War.

*Barnes Wallis.*

# Guy Gibson—Part 1

Wing Commander Guy Penrose Gibson, VC, DSO & Bar, DFC & Bar (12 August 1918 – 19 September 1944), was the first Commanding Officer of the Royal Air Force's No. 617 Squadron, which he led in the 'Dam Busters' raid (Operation Chastise) in 1943, resulting in the destruction of two large dams in the Ruhr area of Germany. He was awarded the Victoria Cross, and in June 1943 became the most highly decorated serviceman in the country. He completed over 170 war operations before dying in action at the age of 26.

Gibson was born in Simla, India, the son of Alexander James Gibson and his wife Leonora (Nora) Mary Gibson. At the time of his birth, his father was an officer in the Imperial Indian Forestry Service, becoming the Chief Conservator of Forests for the Simla Hill States in 1922. In 1924, when he was six, his parents separated. His mother was granted custody of Gibson, his elder brother Alexander and sister Joan, and decided to return to England.

As her family came from Porthleven, Cornwall, she settled first in Penzance. Gibson started school in England at the same school as his sister, West Cornwall College. His mother then moved to London and he was sent as a boarder to Earl's Avenue School, a preparatory school, later known as St George's, in Folkestone, Kent.

In 1932 he started at St Edward's School, Oxford, the same school as Douglas Bader where he was also placed in the same house, Cowell's. Gibson's housemaster was A. F. 'Freddie' Yorke who became Gibson's guardian.

Following her return from India, his mother developed a drinking problem which escalated into alcoholism. Her behaviour became increasingly erratic and sometimes violent towards her children. The school organised lodgings for Gibson and his brother during the school holidays. Nora's younger sister, Mrs Beatrice Christopher, gave Gibson his own room at her house. Her husband, John, helped Nora out with school fees. They also both attended some school functions to support their nephews.

Gibson was an average student academically and played for the Rugby Second XV. His interests included science and photography. At one stage as a teenager, he seems to have become interested and quite expert in the workings of cinema organs. He read all kinds of books, especially the Arthurian legends and Shakespeare. His favourite play was *Henry V*. He was made a house prefect.

From an early age Gibson wanted to fly. He had a picture of his boyhood hero, Albert Ball, VC, the First World War flying ace, on his bedroom wall at his aunt's house. His ambition was to become a civilian test pilot. He wrote for advice to Vickers, receiving a reply from their chief test pilot, Captain Joseph Summers, who wrote that Gibson should first learn to fly by joining the RAF on a short service commission. Gibson applied to the RAF, but was rejected when he failed the Medical Board; the probable reason that his legs were too short. His later application was successful, and his personal file included the remark 'satisfactory leg length test carried out'. He commenced a short service commission in November 1936.

Gibson commenced his flying training on 16 November 1936 at the Bristol Flying School, Yatesbury, with No. 6 Flying Training Course and with civilian instructors. Owing to poor weather the course did not conclude until 1 January 1937. After some leave, he then moved to No. 24 (Training) Group at RAF Uxbridge for his RAF basic training. He was commissioned with the rank of acting pilot officer with effect from 31 January 1937. He then underwent further flying training as a member of the junior section of No. 5 Flying Training Course at 6 Flying Training School, RAF Netheravon. He was awarded his pilot's wings on 24 May 1937.

As part of the Advanced Training Squadron, during summer 1937, he participated in further training at No. 3 Armament

Training Station, Sutton Bridge, Lincolnshire. He opted for bombers as these gave experience in multi-engined planes, this being typical for individuals planning on a civilian flying career. He returned to Netheravon and graduated on 31 August 1937. He passed all his ground exams first time, with an average of 77.29% and a flying rating of 'average'. However, his rating as a companion was below average owing to his sometimes rude and condescending behaviour towards junior ranks and ground crews in particular.

*Guy Penrose Gibson VC.*

### No. 83 (Bomber) Squadron

Gibson's initial posting was to No. 83 (Bomber) Squadron, stationed at RAF Turnhouse, west of Edinburgh. He was assigned to 'A' Flight and was placed under the supervision of Pilot Officer Anthony 'Oscar' Bridgman. The Squadron was flying Hawker Hinds. He joined a settled group of officers from similar minor public school backgrounds. As some stayed with the squadron for a few years, promotion was slow. He was promoted to pilot officer on 16 November

1937. His behaviour towards the ground crews continued to be perceived as unsatisfactory and they gave him the nickname the 'Bumptious Bastard'.

In March 1938, the Squadron was transferred from No. 2 Group to No. 5 Group and relocated to RAF Scampton. In June they moved to RAF Leuchars for an armaments training camp. From October the squadron started their conversion to the Handley Page Hampden, which was completed by January 1939. At a Court of Inquiry in October 1938, Gibson was found guilty of negligence after a taxiing incident at RAF Hemswell. He spent Christmas Day 1938 in hospital at RAF Rauceby with chickenpox. He was then sent on convalescent leave, returning to the squadron in late January.

In Spring 1939 the squadron took part in an armaments training camp at RAF Evanton near Invergordon in Scotland. With the likelihood of war increasing and as part of a plan to improve standards, Gibson was sent on a navigation course at Hamble near Southampton. He did not appear to take the course seriously, but passed with an average mark. The instructor added the comment 'could do well'. He was due to leave the RAF, but was retained owing to the outbreak of hostilities in Abyssinia. In June he was promoted to flying officer. On 25 July the squadron made a long-distance flight to the south of France. They participated in Home Defence exercises over London in August. He then went on his summer leave. At this stage of his career, he had never flown or landed a plane at night.

### First operational tour: No. 83 Squadron, Bomber Command

Gibson was recalled from leave back to Scampton by telegram on 31 August 1939. Gibson flew on 3 September 1939, two days after the start of the Second World War. He was one of the pilots selected to attack the German fleet, which was near Wilhelmshaven. He took off at 1815 hours. The operation was aborted owing to bad weather and he landed back at RAF Scampton around 2300 hours.

On 5 September while in the Mess, he was bitten by a dog. His arm was put in a sling and he was granted 36 hours leave. This allowed him to attend his brother's wedding in Rugby, where he was Alick's best man. On his return, the squadron had moved to Ringway near Manchester under the Scatter Scheme. They were there for 10 days. The squadron did not fly on another operation until December, during the Phoney War.

In February 1940, Gibson was one of the members of the squadron put on temporary secondment to Coastal Command at RAF Lossiemouth. On 27 February, he participated in an operation that was sent to attack a U-Boat. However, owing to various communications problems, one of the aircraft dropped its bombs on a Royal Navy submarine. The senior officers involved with the incident were censured for their failure to bring the squadron up to a satisfactory standard. The squadron then underwent a period of intensive training.

The period from April to September 1940 was one of the most operationally intense periods of Gibson's career. He completed 34 operations in 5 months, with 10 in June. The type of operation varied from 'gardening'—laying mines in various seaways and harbour entrances—to attacks on capital ships, as well as attacks on ground-based military and economic targets. During this time, he acquired a reputation for being seemingly fearless, particularly as he was willing to fly in marginal weather. He was awarded the Distinguished Flying Cross (DFC) on 9 July 1940. He was trained for a low-level attack on the Dortmund-Ems canal, but he missed the actual raid on 12 August. On his return from a raid on Lorient on 27 August, he spotted a Dornier Do 215 and attacked it. He was credited with a 'probable' kill. He was promoted to flight lieutenant on 3 September 1940. His last operation with the squadron was to Berlin on 23 September 1940. Arthur Harris, then the Air Officer Commanding (AOC) No. 5 Group, later described Gibson as the 'most full-out fighting pilot' under his command at this time.

As was usual practice, to give pilots a rest from operations, Gibson was posted as a flying instructor to No. 14 Operational Training Unit (OTU) at RAF Cottesmore. He was there for two weeks, part of which was spent on leave, but he did not settle. He was then transferred to No. 16 OTU at RAF Upper Heyford. Meanwhile, Air Marshal Sholto Douglas, Deputy Chief of the Air Staff, and Air Vice Marshal Trafford Leigh-Mallory, AOC No. 12 (Fighter) Group, made an appeal to Harris for bomber pilots with their night-flying experience to fly night fighters. Gibson volunteered. Harris wrote a letter introducing the pilots, which included the comment 'a hand-picked bunch of which Gibson is the best'. Harris agreed to help Gibson's career when he had completed this tour with 'the best command within my power'.

## Second operational tour: No. 29 Squadron, Fighter Command

Gibson was ordered to report on 13 November 1940 to No. 29 Squadron as the commander of 'A' Flight. The squadron was stationed at RAF Digby, but flew from a small satellite field at RAF Wellingore about six miles away. The Officers' Mess was nearby in The Grange. When he arrived, the Commanding Officer, Squadron Leader Charles Widdows, was in the process of rebuilding the squadron following an outbreak of indiscipline that nearly led to its disbandment during July 1940. He was weeding out under-performing pilots and replacing his flight commanders. Gibson attracted some hostility from some longer-standing members of the squadron. because as one of these new flight commanders, he was seen as part of Widdows' reforms and he had been chosen over an existing member of the squadron. He had also come from a Bomber squadron. The root cause of the low morale was a lack of combat success. The Bristol Blenheim was not designed as a night fighter and the airborne interception was still in its very early days of development. Also, Widdows was required to split the squadron up with a few pilots each at Ternhill, Kirton and Wittering and with no more than half at Digby at any one time. Gibson flew six operations in Blenheims.

The squadron started to convert to the Bristol Beaufighter I and Widdows personally supervised his pilots during their conversion. Gibson's first flight in a Beaufighter was on 1 December 1940. He then undertook some intensive training on AI procedure. He found the night-fighter culture very different from bombers as the two-man crew had to work as a team with the pilot relying on the guidance of the AI operator to find their targets. Gibson made his first operational flight in a Beaufighter on 10 December with Sergeant Taylor as his AI operator. That winter saw bad weather and he flew only three operations in the whole of January. He claimed a kill on 12 March, but it was not confirmed. However, his kill on 14 March was confirmed as a Heinkel He 111. He went to Skegness to collect the tail assembly as trophy for the squadron and the crew's dinghy for himself. He was attacked by an intruder when landing at Wellingore on 8 April. Gibson was unharmed, but his AI operator, Sergeant Bell, was injured in the leg.

In April, Widdows obtained a transfer for the squadron from 12 to 11 Group and a move to RAF West Malling in Kent. Gibson flew down with him on 25 April to inspect the facilities. The full squadron flew down on 29 April. Gibson was promoted to acting squadron leader towards the end of June 1941 and started to deputise for the commander in his absence. Widdows was promoted to station command and was replaced by Wing Commander Edward Colbeck-Welch. Gibson claimed two more kills which were confirmed. Another unidentified bomber, possibly a Heinkel, was claimed in flames on 3/4 May. On 6 July he downed a Heinkel He 111H-5 of 8/KG4 near Sheerness. His AI operator on all his successful claims was Sergeant R.H. James, who was awarded a Distinguished Flying Medal. However, the Luftwaffe's bombing offensive was tailing off and Gibson started to get bored by the relative safety, and began to describe patrols as 'stooge patrols' in his log book. He made some further interceptions but his guns or cannons failed. He was also concerned by his relative lack of success compared with his fellow flight commander Bob Braham. He seems to have been happy at West Malling and said 'Of all the airfields in Great Britain, here, many say, including myself, we have the most pleasant'. His final patrols with the squadron were flown on 15 December. He left with both flying and gunnery ratings of above average. He was awarded a Bar to his DFC.

Again, as a rest from operations, Gibson was due to be posted to an OTU, this time No. 51 OTU, RAF Cranfield as Chief Flying Instructor. By now he had decided he wanted to return to bombers. Despite a visit to HQ No. 5 Group on 15 December to petition for a transfer, Fighter Command insisted he had to go to Cranfield. His opportunity came a few weeks later when on 22 February 1942, Arthur Harris was appointed Air Officer Commanding-in-Chief of Bomber Command. Harris fulfilled his promise made in September 1940. He called Gibson for an interview. On 22 March, Harris wrote to Air Vice Marshal John Slessor, AOC No. 5 Group, explaining his intention to promote Gibson to acting wing commander to put him in command of a Lancaster squadron. Harris suggested No. 207 Squadron. Slessor exercised his discretion and appointed Gibson CO of No. 106 Squadron. Gibson was posted from No. 51 OTU and sent on leave until April, which he spent in south Wales.

Third operational tour: No. 106 Squadron, Bomber Command

When the newly promoted Wing Commander Gibson joined No. 106 Squadron at RAF Coningsby, morale was good, but there was serious disappointment with the new twin-engined Avro Manchester because its Rolls Royce Vulture engines were unreliable. Therefore, the squadron was scheduled to convert to the four-engined Avro Lancaster, equipped with Rolls-Royce Merlin engines as soon as they became available.

Gibson eased himself back into bomber operational flying with a mine-laying operation in the Baltic on 22 April 1942 and completed three more sorties in the Manchester during the following 3 weeks.

April 1942 was a good month for the squadron. They flew on eighteen nights, six consecutively and the improvements in performance were noted by analysts at both No. 5 Group and Bomber Command. The Lancasters started to arrive during May and an ad hoc training plan was started while normal operations were maintained. Gibson made his first flight in a Lancaster in early May.

As a commander, Gibson's main concern was to be seen to share the risk. He continued to show unremitting aggression with a selectivity towards harder targets rather than easier ones. He expected the same determination from everyone on the squadron. He was ruthless in screening crews for reliability. The station's Medical Officer became expert in determining which crews were simply unlucky in contrast with genuine malingerers. However, he was capable of serious misjudgements on occasions, and could be prone to unreasonable outbursts and the persecution of some crews and their members.

Like Widdows, he carefully supervised new crews and eased them into operational flying with 'Nasturtium training' — mine-laying and then easier targets. He was pressurised to expose them earlier to greater risks and he acquired a reputation for not accepting any interference in how he ran the squadron.

Gibson's exercise of summary discipline tended towards constructive tasks aimed at improving the efficiency of the squadron such as maintenance of aircraft, engines or weapons. He was responsible for the emergence of an inner circle of officers who shared his intensity for operations. Their off-duty activities included swimming, water polo and shooting. However, his behaviour towards NCOs and ground crews could still be a problem. Soon after his arrival, the NCOs perceived one incident he was involved in with them as particularly high-handed and the ground crews quickly gave him the nickname 'The Boy Emperor'.

On 11 May, he was hospitalised at RAF Rauceby. The exact reason is unknown, but suggestions include a sinus or middle ear problem. He was then sent on two weeks convalescent leave. This absence meant he was unable to participate in Operation Millennium, the '1,000 Bomber raids', the first of which was made on Cologne on 30 May 1942. He found this frustrating because this raid saw the introduction of the Bomber stream. This was where the aircraft were concentrated together in an attempt to overwhelm the defences, with each allocated a specific place, height band, and time slot. This period saw the introduction of aiming-point photography. Gibson tried it out and then encouraged all aircrews to become 'photo minded'. Obtaining good aiming point photographs quickly became a competition both within, and between squadrons.

On his return he continued to build up his experience with the Lancaster. He flew with his friend, pilot John Hopgood on 4 July and then on a long cross-country flight the day after, 5 July. He made his first operational flight in a Lancaster on 8 July with Dave Shannon as his second pilot. They were together again on 11 July when they went to Danzig. They were appalled when they were sent on a daylight Mohling raid to the Krupps in Essen on 18 July. It was known as a difficult and dangerous target at night and they were relieved to be recalled when near Vlissingen. They jettisoned their bombs over the sea before returning.

The squadron was selected for special training in the use of two kinds of new bombsight for use with a special bomb designed for attacks on capital ships. However, Gibson advised that the aircraft should not attack any ships below 8,500 feet. They put this training into practice with a marathon flight to Gdynia on 27 August 1942. The targets were *Gneisenau* and *Scharnhorst*. Gibson again flew with Shannon and they swapped places during the flight. There was significant unexpected haze over the target when they arrived. Gibson's bomb aimer, Squadron Leader Richardson, a bombing instructor from RAF Manby, requested twelve practice runs over the target, but they still failed to damage the ship. In fact, no ships were damaged during the raid, but the squadron's preparation for the raid was noted by Harris and Air Commodore Alec Coryton, the AOC No. 5 Group.

On 30 September the squadron moved from Coningsby to RAF Syerston in Nottinghamshire. They expected this move to be only temporary while the runways were concreted, but problems at Coningsby meant it became permanent.

Gibson quickly formed a good relationship with RAF Syerston's station commander Group Captain 'Gus' Walker. In October, they were required to conduct low-level training exercises with aircraft flying in formations of threes and sixes. This training was put to use in a raid on the 17th on Le Creusot in France. Gibson and Hopgood were among the pilots sent to attack the electric transformer station at nearby Montchanin. Later in the month they started to attack Italian targets including Genoa, Milan and Turin. In November 1942 Gibson was awarded the Distinguished Service Order (DSO).

On 8 December Gibson did not fly. He was in the control room with Walker watching the aircraft taxiing for take-off. Walker noticed some incendiaries which had fallen out of the bomb bay of a reserve Lancaster located near the main bomb dump. The incendiaries had ignited. Walker drove out to the plane and tried to move the incendiaries with a rake. He lost his arm in the subsequent explosion of the 4,000 lb 'cookie' bomb still in the aircraft's bomb bay. He was replaced by Group Captain Bussell.

On 16 January 1943, Gibson took the BBC's war correspondent, Major Richard Dimbleby on a sortie to Berlin. Dimbleby described the raid in a later radio broadcast. Gibson was very pleased with the outcome, as he always wanted to communicate what life was like for the aircrews. On 12 March, he made his final flight with the squadron to Stuttgart. He flew on three engines and was forced to stay low throughout the raid.

Bussell recommended Gibson for a Bar to his DSO, but this was reduced to a second Bar to his DFC at HQ No. 5 Group owing to the recent award of the DSO. However, Harris confirmed the Bar to Gibson's DSO with the comment 'any Captain who completes 172 sorties in outstanding manner is worth two DSOs if not a VC. Bar to DSO approved'. Gibson was informed on 25 March, after he left the squadron. Gibson was expecting to go on leave to Cornwall and was therefore shocked when he received a call from HQ No. 5 Group to inform him he was being posted there to write a book.

**No. 617 Squadron and Operation Chastise**

**Formation of Squadron X**

After the decision was made to attack the Ruhr dams, Harris decided to hand the direct responsibility for the detailed planning, preparation and execution to Air Vice Marshal Ralph Cochrane, AOC No. 5 Group. Harris told him he must form a new squadron and nominated Gibson as the CO.

On 18 March Gibson attended an interview at HQ No. 5 Group where Cochrane asked him if he was willing to fly on 'one more trip'. Gibson indicated that he was. He attended a further interview the following day when he was told that he was to command a new squadron, which would be required to fly low at night with an objective that had to be achieved by 19 May. At this meeting, he was introduced to Group Captain John Whitworth, the commander of RAF Scampton where the new squadron was to be stationed.

### Selection of aircrew

Unusually, Gibson was allowed to select crews for the new squadron with help from Group Captain Satterly, the Senior Air Staff Officer (SASO) No. 5 Group. The initial intention was for the squadron to be formed from volunteer tour-expired crews. Bomber Command then stated they should have completed or nearly completed two tours of operations.

He selected Squadron Leaders Maudslay and Young as his flight commanders. His selection of Young resulted in the transfer of the whole of 'C' Flight from No. 57 Squadron into the new one. Some crews or pilots were known to him including Hopgood and Shannon, who by this time had transferred from No. 106 squadron to the Pathfinders and No. 83 Squadron. He selected Harold 'Mick' Martin for his low-flying expertise. Of Gibson's regular crew from No. 106 Squadron, only Hutchinson, the wireless operator, volunteered for this new one. He brought in a friend, Taerum, a Canadian navigator. Taerum in turn brought in another friend, Spafford, a bomb-aimer and they joined Hutchinson in Gibson's crew. In the end some crews had not completed one tour, with some individuals having flown fewer than ten operations.

Gibson was strict in screening the crews during training. That not all the crews were known to him is reflected in how two crews were posted off the squadron as not satisfactory and another crew chose to leave after their navigator was deemed unsatisfactory.

### Training of No. 617 Squadron

Gibson arrived at Scampton on 21 March. His office was on the 1st floor in No. 2 Hangar. His immediate task was to get the general administration organised. He delegated this and the adjutant assigned from No. 57 Squadron was quickly replaced with Flight Lieutenant Humphreys from RAF Syerston. Humphreys was instrumental in the rapid establishment of the squadron. The ground staff started to muster from 21 March and were fully present by 27 March. Flight Sergeant Powell inspected them and weeded out those he felt other squadrons had off-loaded. The aircrews started to arrive from 24 March.

On 24 March Gibson travelled to Burhill near Weybridge for his first meeting with Barnes Wallis. Wallis discovered Gibson had not been cleared for a full briefing and therefore could not be told the targets. Wallis was able to explain the design and operation of the new weapon, Upkeep and showed him films from its trials. It was a depth charge which, if rotated with backspin and dropped at the correct speed and altitude, would bounce across the surface of a body of water towards a target. This bouncing behaviour gave it its nickname the bouncing bomb. The crews usually referred to it as a mine.

On 27 March Group Captain Satterley provided Gibson with 'most secret' written orders, including a description of the attack and the general plan for the squadron's preliminary training. From these Gibson learnt that the targets were 'lightly defended special targets' which reduced his suspicion that they were training to attack the *Tirpitz*. The orders included a list of nine lakes and reservoirs in the Midlands and North Wales, for training flights and target practice. They included Eyebrook Reservoir, near Uppingham, Rutland, Abberton Reservoir near Colchester and Derwent Reservoir and some of the earliest flights made by the new No. 617 Squadron, were reconnaissance flights over these bodies of water. A recommendation to maximize the training time available was to use simulated night flying as developed by the USAAF. This required the cockpit to be covered in blue celluloid and the pilots and bomb aimers to wear goggles with amber-tinted lenses. Gibson wanted six aeroplanes converted but only two became available, the first on 11 April.

Another important factor was the need for a specially adapted version of the Lancaster, the B.III (Special), officially the 'Type 464 (Provisioning)'. The bomb bay doors were removed and the ends of the bay were covered with fairings. Upkeep was suspended on pivoted, vee-shaped struts which sprang apart when the bomb-release button was pressed. A drive belt and pulley to rotate the bomb at 500 rpm was mounted on the starboard strut and driven by a hydraulic motor housed in the forward fairing. The mid-upper turret was removed and a more bulbous bomb aimer's blister was fitted. The first adapted aircraft arrived at Scampton on 8 April.

On 28 March, Gibson made his first flight to explore the low-flying requirement. He took Hopgood and Young with him and found low flying during daylight satisfactory but during an attempt at dusk the difficulty of their task became apparent, when they nearly ditched. On 29 March, Gibson was shown scale models of the Möhne and Sorpe dams by Cochrane at HQ 5 Group. He then attended a further meeting with Wallis at Weybridge. At this meeting he rejected Wallis's proposal of a daylight raid.

The squadron commenced daily flying training at the beginning of April with long cross-country flights with precise turning points to develop their navigation skills. They then started to practise low flying over water. The squadron completed over a thousand flying hours by the end of April and Gibson was able to report to Whitworth, that they could fly pinpoint to pinpoint at low level at night, could bomb using a rangefinder and fly over water at 150 ft (46 m). On 24 April Wallis made a request for the altitude to be reduced to 60 ft (18 m). Gibson reported on 27 April that it was possible and the training was adapted accordingly.

Gibson was closely involved with discussions about the design, trial and approval of the solutions developed for the various technical issues encountered. These included the Dann bomb sight and the 'Spotlight Altimeter Calibrator', which was the name given to the spotlights attached to the Lancasters, to ensure the determination of the correct height above a body of water. Security was Gibson's constant concern and he was especially displeased to learn from his bombing leader Watson that he had been shown details of the targets within days of his arrival at RAF Manston. Gibson wrote to Cochrane who raised his concerns about this 'criminal' breach of security at the highest levels.

From the beginning of May squadron training shifted to the tactical aspects of the operation. On 1 May Gibson communicated to Wallis his confidence that the operation would succeed. He repeated this optimism in his weekly report to Whitworth on 4 May where he described the squadron as 'ready to operate'. On 6 May he held a conference with the pilots to explain the tactical aspects. They flew a rehearsal that evening with Gibson directing a group by radio telephony (R/T) on the spot over the Eyebrook and Abberton Reservoirs. A second group went to the Derwent Reservoir and a third to the Wash. On 10 May, Satterly sent the draft handwritten operation order to Whitworth for review and revision, to be returned by 1600 hours on 12 May. It included, how the squadron would be split into waves to attack the targets, reserves, likely defences and exit routes; Gibson provided detailed comments. Despite Gibson's confidence, there still had not been a successful release of a live Upkeep, which took until 11 May. Most of the crews were able to practise at Reculver from 11 to 14 May. Gibson practised at Reculver in Lancaster ED932/AJ-G, the aircraft he used on the raid. The aircraft's call letters were the same as his father's initials: AJG. On 14 May the squadron flew on a full-dress rehearsal designed to simulate the routes, targets and the geography of the raid. Gibson took Whitworth with him and described the outcome in his log book as 'completely successful'.

Cochrane travelled to Scampton on 15 May to inform Whitworth and Gibson that the operation would take place the following evening, over 16/17 May. At about 1600 hours, Gibson travelled with Cochrane on his return to Grantham. Here he discussed the draft operation order with Satterly and Wing Commander Dunn, No. 5 Group's chief signals officer. He returned to Scampton and at 1800 hours at Whitworth's house, along with Wallis, he briefed Young and Maudsley, his flight commanders, and Hopgood, the deputy leader, and Hay, the squadron's bombing leader. He had obtained Cochrane's verbal agreement for Hopgood and Hay to attend, which proved beneficial as Hopgood was able to point out the new defences at Huls. After the meeting broke up, Whitworth informed Gibson that his dog had been killed in a road accident. It did not seem to affect Gibson outwardly. He was aware how superstitious some aircrew could be, as the dog was the squadron's mascot. Wallis feared it was a dreadful omen.

# 617 Squadron

As the plans to implement the 'dams raid' progressed, a squadron would be needed to fly the mission. At the time 5 Group was the only one flying Lancaster bombers. The squadron would have to come from within this group, as they were they only one with Lancaster experience. Finding a squadron from within the group would be a problem. Harris was reluctant to withdraw a whole Lancaster squadron from the main force for the duration of the dams mission. A new squadron would have to be formed made up of tour-served crews who would be coming off ops having completed their set number of missions.

Wing Commander Guy Gibson of 106 Squadron was chosen to form and lead this new squadron because of his formidable operational record and reputation for seeing through a task due to his leadership skills and strict discipline. He was asked to do this one more operation after completing his tour with 106 Squadron. He agreed to undertake the operation with no idea of what was being asked.

Unusually, Gibson was granted the authority to pick his own crews. They would have to be experienced veterans who had completed or nearly completed two tours. However, although many believe that 617 Squadron was formed from the very best, highly decorated pilots and aircrew in the allied force, this was far from the truth. The majority of the squadron had no decorations at all and instead of having flown two tours, some were only one third of the way through their first tour. Gibson personally knew very few of the men including his own crew. Only Flight Lieutenant Bob Hutchinson, a radio operator had flown regularly with Gibson at 106 Squadron. They had finished their second tour together. The new squadron would share RAF Scampton with resident 57 Squadron.

In late March 1943 Les and his crew were offered the chance to transfer into a new squadron being formed at nearby RAF Scampton for a 'special' mission. They took a joint decision to transfer together, as Bob Kellow, later explained—'The offer presented to us sounded interesting and with our faith in each member's ability we made up our minds there and then that we would accept the offer and move over as a crew to this new squadron.'

The crew's faith was probably because they had together recognised that Knight was an exceptional pilot, even though he couldn't ride a bicycle or drive a car. Like his colleagues, Henry O'Brien—the rear gunner—admired Les who he regarded as 'the coolest and quickest thinking person I have ever met. And, in my opinion, the most knowledgeable person in the Squadron with respect to his job.'

Les and his crew marched into the newly formed squadron on 25 March 1943. At this time, they had yet to be given a squadron number and they were known as Squadron 'X'. As crews from across the UK assembled there must have been a growing buzz about what they had been brought together for.

Now that the squadron had been formed, training could begin. This was crucial, as the crews would have to learn to fly to extreme limits in order to carry out the attacks on the dams. The new squadron began training for the mission almost immediately. They had very little time to prepare. Learning low flying both day and night was the first and most important task for the crews.

Although it may seem unlikely for experienced air crews, air sickness was the first problem many of them had with low flying. Flying at low level caused intensive turbulent shaking of the aircraft and many of the crews who were used to operating at the rather smoother altitude of 10,000 feet experienced it. Medicine was handed out to those who needed it but some had to be rejected and replaced.

The flying was very intensive; night after night they practiced at first in borrowed Lancasters and later in the modified types as they came through from Avro. In order to make conditions as realistic as possible, they were told to fly over three main locations in England. The Eyebrook Reservoir at Uppingham in Leicestershire, the Abberton Reservoir near Colchester and the Derwent Reservoir near Sheffield. It is important to remember that neither Gibson nor the crews were aware of their targets at this time, the information was absolute top secret and very few people knew. The crews were however beginning to guess what their target may be. At first the rumours were that the target was the German battleship Tirpitz.

Although Gibson was not told officially the target, he was given a very good idea of what he was up against at a meeting with Wallis on 24 March 1943. This was the first time the two men had met. Wallis could not tell Gibson specific details of the mission as he was not on the list of people with clearance for a full briefing, he did however tell him as much as he could. After the meeting, Gibson left in the certain knowledge that his aircraft must attack the targets at a speed of 240 mph at a height of 150 feet, any variation on this and the plans simply would not work.

By the end of March, Squadron 'X' was officially designated the number 617. They were now officially 617 Squadron.

## No. 617 Squadron RAF

No. 617 Squadron was formed under great secrecy at RAF Scampton during the Second World War on 21 March 1943. It included Royal Canadian Air Force, Royal Australian Air Force and Royal New Zealand Air Force personnel and was formed for the specific task of attacking three major dams that contributed water and power to the Ruhr industrial region in Germany: the Möhne, Eder and Sorpe.

The plan was given the codename 'Operation Chastise' and carried out on 17 May 1943. The squadron had to develop the tactics to deploy Barnes Wallis's 'Bouncing bomb', and undertook some of its training over the dams of the Upper Derwent Valley in Derbyshire, as the towers on the dam walls were similar to those to be found on some of the target dams in Germany.

The original commander of 617 Squadron, Wing Commander Guy Gibson, was awarded the Victoria Cross for his part in the raid. After the raid, Gibson was withdrawn from flying (due to the high number of raids he had been on) and went on a publicity tour. George Holden became Commanding Officer in July, but he was shot down and killed on his fourth mission, 'Operation Garlic' in September 1943, in an attack on the Dortmund-Ems Canal. H.B. 'Mick' Martin took command temporarily, before Leonard Cheshire took over as Commanding Officer. Cheshire developed and personally took part in the special target marking techniques required, which went far beyond the precision delivered by the standard Pathfinder units—by the end he was marking the targets from a Mustang fighter. He was also awarded the Victoria Cross.

On 15 July 1943 12 aircraft of the squadron took off from RAF Scampton to attack targets in Northern Italy. All aircraft attacked and proceeded to North Africa without loss. The targets were San Polo d'Enza and Arquata Scrivia power stations; it was hoped that the attacks would delay German troops who were travelling down into Italy on the electrified railway system to support the Italian front. The operation met little opposition but the targets were obscured by valley haze and were not destroyed. The 12 crews returned to Scampton on 25 July from North Africa after bombing Leghorn docks on the return journey. The raid on Leghorn Docks was not a great success, due to mist shrouding the target. On 29 July 1943 nine aircraft took off from Scampton to drop leaflets on Milan, Bologna, Genoa and Turin in Italy. All aircraft completed the mission and landed safely in Blida North Africa. Seven of the aircraft returned to Scampton on 1st August, one on the 5th and the last on the 8th.

Throughout the rest of the war, the squadron continued in a specialist and precision-bombing role, including the use of the enormous 'Tallboy' and 'Grand Slam' ground-penetrating earthquake bombs, on targets such as concrete U-boat shelters and bridges. Several failed attempts were made on the Dortmund-Ems Canal in 1943 ('Operation Garlic') however it was finally breached with 'Tallboys' in September 1944.

A particularly notable series of attacks caused the disabling and sinking of *Tirpitz*. *Tirpitz* had been moved into a fjord in northern Norway where she threatened the Arctic convoys and was too far north to be attacked by air from the UK. She had already been damaged by an attack by Royal Navy midget submarines and a series of attacks from carrier-borne aircraft of the Fleet Air Arm, but both attacks had failed to sink her. The task was given to No. 9 and No. 617 Squadrons where they were deployed to Yagodnik, near Archangel a staging base in Russia to attack *Tirpitz* with Tallboy bombs.

On 15 September 1944, the RAF bombers struck the battleship in the forecastle, which rendered her unseaworthy, so she was sent to the Tromsø fjord where temporary repairs were made so she was anchored as a floating battery. This fjord was in range of bombers operating from Scotland and from there, in October, she was attacked again, but cloud cover thwarted the attack. Finally, on 12 November 1944, the two squadrons attacked *Tirpitz*. The first bombs missed their target, but following aircraft scored two direct hits in quick succession. Within ten minutes of the first bomb hitting the *Tirpitz*, she suffered a magazine explosion at her 'C' turret and capsized killing 1,000 of her 1,700 crew. All three RAF attacks on *Tirpitz* were led by Wing Commander J. B. 'Willy' Tait, who had succeeded Cheshire as Commanding Officer of No. 617 Squadron in July 1944.

Among those pilots participating in the raids was Flight Lieutenant John Leavitt, an American who piloted one of the 31 Lancasters. Leavitt's aircraft dropped one of the bombs that hit Tirpitz dead centre. Despite both squadrons claiming that it was their bombs that actually sank the *Tirpitz*, it was the Tallboy bomb, dropped from a No. 9 Sqn Lancaster WS-Y (LM220) piloted by Flying Officer Dougie Tweddle that is attributed to the sinking of the warship. Flying Officer Tweddle was awarded the Distinguished Flying Cross for his part in the operations against *Tirpitz*. During the Second World War the Squadron carried out 1,599 operational sorties with the loss of 32 aircraft.

*The 617 Squadron RAF Crest.*

# RAF Scampton

Royal Air Force Scampton, or RAF Scampton, is a Royal Air Force station located adjacent to the A15 road near to the village of Scampton, Lincolnshire, and 6 miles (9.7 km) north west of the county town, Lincoln, England.

RAF Scampton stands on the site of a First World War Royal Flying Corps landing field, which had been called Brattleby. The station was closed and returned to agriculture following the First World War being reactivated in the 1930s.

At the outbreak of the Second World War Scampton transferred to No. 5 Group RAF in Bomber Command, playing host to the Hampdens of 49 Sqn and 83 Sqn. On 3 September 1939, six hours after the declaration of war, RAF Scampton launched the first offensive by the Royal Air Force when six Hampdens of 83 Sqn, led by (the then) Flying Officer Guy Gibson and three 49 Sqn Hampdens, one piloted by Flying Officer Roderick Learoyd, were despatched to conduct a sweep off Wilhelmshaven.

Further operations involving Scampton's squadrons concerned them with the hazardous task of low-level minelaying (code named 'Gardening') and the bombing of ships. Scampton squadrons were also involved during the critical stages of the late summer and early autumn of 1940, attacking barges in the channel ports which were being assembled as part of the invasion fleet.

For a short time, the station was home to the Avro Manchester with 49 Sqn and 83 Sqn operating the type. This was a brief liaison with the squadrons subsequently converting to the Avro Lancaster. Forming 83 Conversion Flight (CF) on 11 April 1942, which in turn was followed by 49 CF on 16 May, both squadrons were fully equipped with the Lancaster by the end of June. It was during this period that 83 Sqn took delivery of Lancaster Mk I R5868 which would later become the Station's gate guardian.

*The front entrance to RAF Scampton.*

In turn both resident squadrons were then replaced at Scampton by 57 Sqn. The first departure was that of 83 Sqn which left in August 1942, transferring to RAF Wyton in order to become part of the fledgling Pathfinder Force. This departure resulted in 83 CF moving to RAF Wigsley where it was disbanded into 1654 Heavy Conversion Unit. On 2 January 1943, 49 Sqn departed for RAF Fiskerton with 49 CU disbanding, subsequently becoming 'C' Flight of 1661 Heavy Conversion Unit at RAF Waddington. By early January 1943 this left 57 Sqn as the sole occupier of the base.

Following the development of the Upkeep bouncing bomb, 617 Sqn, originally referred to as 'Squadron X', was formed at Scampton in order to carry out the proposed raid, codenamed Operation Chastise. On the day of the raid, Wing Commander Gibson's dog, Nigger, was run over and killed on the A15 outside the entrance to the base. He was buried later that night, his grave situated outside Gibson's office at No. 3 Hangar.

At the end of August 1943, 57 Sqn and 617 Sqn moved to RAF East Kirkby and RAF Coningsby respectively, so that Scampton's runways could be upgraded. With the increased all up weight of the Lancaster it was apparent

*Nigger's Grave.*

that the load bearing of hardened runways would be required. The airfield closed at the end of August 1943 for the work to take place re-opening in October 1944.

On completion of the required work the area of land which the base occupied had now increased to 580 acres. Following the work control of the station passed from 5 Group to 1 Group with a new arrival following the upgrade being 1690 Bomber Defence Training Flight (BDTF) which arrived on 13 July 1944. The BDTF consisted of Spitfires, Hurricanes and Martinets, the flight undertaking fighter affiliation against

*RAF Scampton Crest.*

bombers. This unit stayed at the station until September 1944, when it moved to RAF Metheringham. Two Lancaster squadrons, 153 Sqn, and later 625 Sqn, of No. 1 Group RAF also arrived at Scampton following the re-opening of the airfield.

The last bombing mission of the Second World War launched from RAF Scampton was on 25 April 1945, when aircraft from 153 Sqn and 625 Sqn were despatched as part of a raid on the Obersalzberg.

RAF Scampton has been operational since the end of the Second World War but the Ministry of Defence has recently announced that the station will close by 2022, with all units relocated elsewhere, and the base be sold off.

# The Bouncing Bomb

A bouncing bomb is a bomb designed to bounce to a target across water in a calculated manner to avoid obstacles such as torpedo nets, and to allow both the bomb's speed on arrival at the target and the timing of its detonation to be pre-determined, in a similar fashion to a regular naval depth charge. The inventor of the first such bomb was the British engineer Barnes Wallis, whose "Upkeep" bouncing bomb was used in the RAF's Operation Chastise of May 1943 to bounce into German dams and explode underwater, with effect similar to the underground detonation of the Grand Slam and Tallboy earthquake bombs, both of which he also invented.

Barnes Wallis' April 1942 paper "Spherical Bomb - Surface Torpedo" described a method of attack in which a weapon would be bounced across water until it struck its target, then sinking to explode underwater, much like a depth charge. Bouncing it across the surface would allow it to be aimed directly at its target, while avoiding underwater defences, as well as some above the surface, and such a weapon would take advantage of the "bubble pulse" effect typical of underwater explosions, greatly increasing its effectiveness: Wallis's paper identified suitable targets as hydro-electric dams "and floating vessels moored in calm waters such as the Norwegian fjords".

Both types of target were already of great interest to the British military when Wallis wrote his paper; German hydro-electric dams had been identified as important bombing targets before the outbreak of World War II, but existing bombs and bombing methods had little effect on them, as torpedo nets protected them from attack by conventional torpedoes and a practical means of destroying them had yet to be devised. In 1942, the British were seeking a means of destroying the German battleship Tirpitz, which posed a threat to Allied shipping in the North Atlantic and had already survived a number of British attempts to destroy it. During this time, the Tirpitz was being kept safe from attack

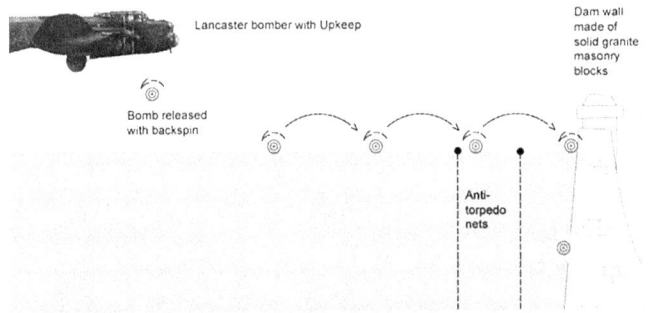

*How the Bouncing Bomb was designed to work.*

by being moored in Norwegian fjords, where it had the effect of a "fleet in being". Consequently, Wallis's proposed weapon attracted attention and underwent active testing and development.

On 24 July 1942, a 'spectacularly successful' demonstration of such a weapon's potential occurred when a redundant dam at Nant-y-Gro, near Rhayader, in Wales, was destroyed by a mine containing 279 pounds (127 kg) of explosive: this was detonated against the dam's side, underwater, in a test undertaken by A.R. Collins, a scientific officer from the Road Research Laboratory, which was then based at Harmondsworth, Middlesex.

A.R. Collins was among a large number of other people besides Barnes Wallis who made wide-ranging contributions to the development of a bouncing bomb and its method of delivery to a target, to the extent that, in a paper published in 1982, Collins himself made it evident that Wallis 'did not play an all-important role in the development of this project and in particular, that very significant contributions were made by, for example, Sir William Glanville, Dr. G. Charlesworth, Dr. A.R. Collins and others of the Road Research Laboratory.' However, the modification of a Vickers Wellington bomber, the design of which Wallis himself had contributed to, for work in early testing of his proposed weapon, has been cited as an example of how Wallis 'would have been the first to acknowledge' the contributions of others. Also, in the words of Eric Allwright, who worked in the Drawing Office for Vickers Armstrong at the time, 'Wallis was trying to do his ordinary job [for Vickers Armstrong] as well as all this – he was out at the Ministry and down to Fort Halstead and everywhere'; Wallis's pressing of his papers, ideas and ongoing developments on relevant authorities helped ensure that development continued; Wallis was principal designer of the models, prototypes and 'live' versions of the weapon; and, perhaps most significantly, it was Wallis who explained the weapon in the final briefing for RAF crews before they set off on Operation Chastise, to use one of his designs in action.

A distinctive feature of the weapon, added in the course of development, was back-spin, which improved the height and stability of its flight and its ability to bounce, and helped the weapon to remain in contact with, or at least close proximity to, its target on arrival. Back-spin is a normal feature in the flight of golf balls, owing to the manner in which they are struck by the club, and it is perhaps for this reason that all forms of the weapon which were developed were known generically as 'golf mines', and some of the spherical prototypes featured dimples.

It was decided in November 1942 to devise a larger version of Wallis's weapon for use against dams, and a smaller one for use against ships: these were code-named 'Upkeep' and 'Highball' respectively. Though each version derived from what was originally envisaged as a spherical bomb, early prototypes for both Upkeep and Highball consisted of a cylindrical bomb within a spherical casing.

Development, testing and use of Upkeep and Highball were to be undertaken simultaneously, since it was important to retain the element of surprise: if one were to be used against a target independently, it was feared that German defences for similar targets would be strengthened, rendering the other useless. However, Upkeep was developed against a deadline, since its maximum effectiveness depended on target dams being as full as possible from seasonal rainfall, and the latest date for this was set at 26 May 1943. In the event, as this date approached, Highball remained in development, whereas development of Upkeep had completed, and the decision was taken to deploy Upkeep independently.

The bomb is dropped close to the surface of the lake. Because it is moving almost horizontally, at high velocity and with backspin, it bounces several times instead of sinking. Each bounce is smaller than the previous one. The 'bomb run' is calculated so that at its final bounce, the bomb will reach close to the target, where it sinks. A hydrostatic pistol causes it to explode at the right depth, creating destructive shockwaves.

Testing of Upkeep prototypes with inert filling was carried out at Chesil Beach, Dorset, flying from RAF Warmwell in December 1942, and at Reculver, Kent, flying from RAF Manston in April and May 1943, at first using a Vickers

Wellington bomber. However, the dimensions and weight of the full-size Upkeep were such that it could only be carried by the largest British bomber available at the time, the Avro Lancaster, and even then, had to undergo considerable modification in order to carry it. In testing, it was found that Upkeep's spherical casing would shatter on impact with water, but that the inner cylinder containing the bomb would continue across the surface of the water much as intended. As a result, Upkeep's spherical casing was eliminated from the design. Development and testing concluded on 13 May 1943 with the dropping of a live, cylindrical Upkeep bomb 5 miles (8 km) out to sea from Broadstairs, Kent, by which time Wallis had specified that the bomb must be dropped at 'precisely' 60 feet (18 m) above the water and 232 miles per hour (373 km/h) groundspeed, with back-spin at 500 rpm: the bomb 'bounced seven times over some 800 yards, sank and detonated'.

Back-spin was to begin 10 minutes before arriving at a target, and was imparted via a belt driven by a Vickers Jassey hydraulic motor mounted forward of the bomb's starboard side. This motor was powered by the hydraulic system normally used by the upper gun turret, which had been removed.[citation needed] Height was checked by a pair of intersecting spotlight beams, which, when converging on the surface of the water, indicated the correct height for the aircraft - a method devised for the raid by Benjamin Lockspeiser of MAP, and distance from the target by a simple, hand-held, triangular device: with one corner held up to the eye, projections on the other two corners would line up with pre-determined points on the target when it was at the correct distance for bomb release. In practice, this could prove awkward to handle, and some aircrews replaced it with their own arrangements, fixed within the aircraft itself, and involving chinagraph and string.

In the operational version of Upkeep, known by its manufacturer as 'Vickers Type 464' the explosive charge was Torpex, originally designed for use as a torpedo explosive, to provide a longer explosive pulse for greater effect against underwater targets; the principal means of detonation was by three hydrostatic pistols, as used in depth charges, set to fire at a depth of 30 feet (9 m); and its overall weight was 9,250 pounds (4,200 kg), of which 6,600 pounds (3,000 kg) was Torpex. Provision was also made for 'self-destruct' detonation by a fuze, armed automatically as the bomb was dropped from the aircraft, and timed to fire after 90 seconds. The bomb was held in place in the aircraft by a pair of calipers, or triangulated carrying arms, which swung away from either end of the bomb to release it.

On the night of 16/17 May 1943, Operation Chastise attacked dams in Germany's Ruhr Valley, using Upkeep. Upkeep was not used again operationally. By the time the war ended, the remaining operational Upkeep bombs had started to deteriorate and were dumped into the North Sea without their detonation devices. In January 1974, under Britain's 'thirty-year rule', secret government files for both Upkeep and Highball were released, although technical details of the weapons had been released in 1963.

*The Upkeep Bouncing Bomb.*

*Testing the Bouncing Bomb.*

*Upkeep mounted on a Lancaster.*

By the end of April 1943, the squadron had flown over 1,000 hours in training and as expected they had encountered special problems with flying at low level and simulating night time flying. The main four problems were: simulating night time flying in daylight; following a map at low level; maintaining a height of 150 feet; and estimating the release point of the weapon from the target to ensure an accurate impact.

The first two problems were relatively easily solved. In order to simulate night flying during the day, blue celluloid was fitted over the cockpit windscreen, side windows and gun turrets while the crew wore amber—tinted goggles. Map reading was made easier by producing strip maps on rollers following each aircraft's particular route.

Height was a much more difficult and crucial problem. Normal Main Force squadrons relied on numbers and mutual defence to confuse and overwhelm enemy radar and defences. However, a small group of bombers flying at normal operational height would soon be either; detected and destroyed by enemy fighters, or successive waves of anti—aircraft batteries from the coast to the target. Not only would the attack have to come at low level, the entire flight there and back would have to be done at low level too.

The problem was that the standard altimeter was useless at low levels such as 150 feet. On 28 March 1943 Gibson with Hopgood and Young on board, had flown over the Derwent Reservoir to see how difficult it was to fly over water at 150 feet with hills all around. During the daylight he had no real problems but when dusk came, he could not distinguish the horizon from the water surface and nearly flew into the water.

The solution to flying at 150 feet was found by the Royal Aircraft Establishment. A year earlier they had been experimenting with spotlights fitted under Hudson bombers in order to gauge their height while attacking U—boats at night. It had not worked very well due to choppy waves in the sea but over a smooth lake it might. After experimenting with it for a while, two lamps were fitted, one in the nose and one behind the bomb bay. They were angled so that the two beams would meet when the aircraft was at exactly 150 feet (46 metres). It would be the job of the navigator to look down through the starboard (right) cockpit window and talk the pilot down until the lamps met at the required altitude.

Both Harris and the crews of 617 Squadron were very shocked at the idea of lighting up an aircraft at the moment it was about to attack when it was at its most vulnerable. Harris was furious 'I will not have aircraft flying about with spotlights on in defended areas' he said angrily. Although he was right, there was no other way to achieve the required altitude. The bombers would have to attack lit up.

As the problems the crews were having with training were slowly being solved and they were becoming good at flying to the limits specified for the attacks. The development of the bouncing bomb was not coming along so smoothly though...

Testing of the bomb was continuing while the crews were on training. Vickers' Mutt Summers and Avro's chief test pilot Sam Brown had been dropping the bombs from the height specified of 150 feet but they continued to break up when first hitting the water.

It is a long living myth that the bomb was a spherical shape when in actual fact it was more like an oil drum (cylindrical). This is mainly due to the 1954 film *The Dambusters*, which portrays the bomb as a spherical shape. This was not entirely inaccurate because Wallis had indeed intended his weapon to be spherical. The reason for the change is outlined below and is due to the bomb breaking apart upon impact with the water. The reason the film kept the bomb as a spherical shape was because at the time it was made, the bomb was still on the secrets list and the film makers were unable to show its actual shape. The bouncing bomb was not removed from the secrets list until 1963 despite the Germans knowing literally everything about it.

When Wallis was experimenting with half—size prototypes, he found that the casing of the bomb which gave it its spherical shape broke upon impact with the water. This was because the casing was made from wood due to the lack of steel available at the time. Despite efforts to increase the strength of the casing it continued to break. Wallis eventually decided to forget about the casing because the cylinder inside the container which contained the explosives often continued and did indeed bounce as he had expected.

However, when Summers and Brown were testing the actual full-size prototypes of the weapon, they totally broke apart upon impact with the water when dropped from 150 feet. The only way to stop them from breaking apart was to reduce the height from which they were dropped. Wallis recalculated and found that 60 feet (18 metres) was the maximum height from which the bombs could be dropped without being torn apart.

With only three weeks to go to the planned date of the raid, Wallis asked Gibson if he and his boys could fly at the perilously low level of 60 feet. Gibson took a deep breath and promised that they would.

The final major problem to be solved was judging the distance at which the bomb had to be released to hit the dam exactly where it needed to be.

Once again, the Royal Aircraft Establishment came up with the solution. They produced simple hand—held Y—shaped wooden sights. The bomb aimer would simply look through an eyepiece at the base of the Y and when the two nails on the arms lined up with the towers on the dam, it was the correct range for releasing the weapon. This was about 400—450 yards (366—411 metres).

In practice, some of the bomb aimers found it very difficult to keep the aimer steady with one hand while the aircraft was bouncing around. Some aimers therefore devised their own means based on the same idea. These ideas generally involved marks on the clear vision panel in the nose and a piece of string to line up with the towers of the dam. It was a very simple idea involving basic mathematics, but it worked!

Communication was going to be critical on the raid. The success of this mission in particular would depend upon not only the teamwork within each aircraft but also the whole squadron.

In a normal bomber stream, aircraft attacked in a conveyor belt system dropping their bombs almost simultaneously on a broad target. This raid relied upon individual precision attacks. Gibson would have to act as the master bomber calling each aircraft in one at a time, essentially co-ordinating the attack and making adjustments as required. Good air to air communication was therefore essential, but the standard R/T radio sets performed badly at low level at night. To solve the problem, they fitted VHF sets normally used by fighter aircraft.

On 11 May 1943, just five days before the night of the attack, the squadron began training with actual bombs at Reculver (although they were not actually filled with explosives). They were amazed to see the drums bouncing over the water right up to the beach. Still they did not know their targets! After seeing the weapon in operation, it reignited talk that the target was the Tirpitz or even U–boats.

In only eight weeks, 617 Squadron was ready to go. Wallis, Chadwick and everyone involved had managed to perfect the bouncing bomb, modify the Lancasters (including designing a release mechanism), form and prepare an entire squadron and iron out all of the major problems facing them. Their effort was already a great achievement.

While on training with real bombs at Reculver, both Shannon and Maudslay damaged their Lancasters by dropping their bombs too low and being caught in a huge column of water thrown up after it hit the water. By the time the attack came five days later, Maudslay's aircraft could not be repaired and the attacking force was down to 19 aircraft from the 20 originally intended by Gibson. Gibson had picked 21 crews for the squadron, the 20 to fly and one reserve. Coincidently, both Divall and Wilson had sickness and their crews would not fly. This therefore left 19 aircraft and 19 crews. The attacking force had been determined out of Gibson's hands.

# The Dams Raid Briefing

With only two days to go before 'Operation Chastise' was due to be launched, the Chiefs of Staff had not confirmed the targets. More worryingly, they had not even made up their minds whether to launch the attack at all. The Navy had great hopes that a variation on the bouncing bomb called 'Highball' that could be used to sink enemy shipping and U—boats and thought that the dams raid would compromise the secrecy of Highball. They did however decide to go for the dams because Highball was still having teething problems and if they waited for the problems to be solved, they would have missed the window in which the lakes were full.

The day before the attack on 15 May 1943, Gibson was given a full briefing on the attack and the targets. Until then he and the squadron had still been kept in the dark. The three main objectives were: Target A—the Möhne Dam; Target B—the Eder Dam; and Target C—the Sorpe Dam. The secondary targets were: Target D—the Lister Dam, Target E—the Ennepe Dam; and Target F—the Diemel Dam. Gibson informed his two flight commanders Young and Maudslay, his deputy leader Hopgood and Bob Hay, Martins' bomb aimer who was the squadron bombing leader.

After the meeting, Gibson was informed that his black Labrador, 'Nigger', had been hit by a car and killed outside RAF Scampton's main gate. Nigger had flown with Gibson on many occasions and had become one of the 'boys' to crews, often enjoying a pint with them. Gibson did not want the others to know in case they thought it was a bad omen. He spent the night in a mood of depression. The next day (16 May) he arranged for the dog to be buried at around midnight, just about the same time he was due to lead the first wave into attack the Möhne Dam.

On 16 May 1943 at 1800 hours crew members assembled for the final briefing. As the last of 133 airmen clambered their way into the briefing room, a door closed behind them, with Service Police preventing any further entry. Opposite the doorway, Air Vice Marshal Ralph Cochrane—the Commanding Officer of 5 Group—sat with other senior officers and Barnes Wallis, who Gibson introduced. Gibson outlined the general plan of attack before the crews were briefed in detail about routes, call signs, codewords, weather conditions and ammunition loads. For many of the crews this was the first time Gibson had spoken to them.

The attacks would be carried out in three waves. The first wave of nine aircraft (AJ–G, AJ–M, AJ–P, AJ–A, AJ–J, AJ–Z, AJ–L, AJ–B and AJ–N) would take off in three sections ten minutes apart. They would fly a southerly route crossing the enemy cost at the Scheldt estuary in Holland. Their first target would be the Möhne Dam. Wallis believed that only one bomb would be required to cause a breach in the dam. The planners allowed Gibson to use three, firstly in case one was not enough and secondly to expand the gap.

Once the Möhne Dam had been breached, the aircraft that had attacked and no longer had a bomb would turn back home while the remaining aircraft with bombs would go onto Target B—the Eder Dam. After breaching the Eder Dam, the process would be repeated and aircraft with remaining bombs would proceed to Target C—the Sorpe Dam.

The Sorpe Dam would be the primary target of the second wave consisting of five aircraft (AJ–T, AJ–E, AJ–W, AJ–K and AJ–H). The second wave would actually leave RAF Scampton first in order to fly a more northerly route to the Dutch island of Vlieland then down the Zuider Zee and join the flight path of the first wave just over the German border. The two routes were chosen to suggest to enemy radar that these were minor attacks. After attacking the Sorpe Dam, the second wave would use any remaining bombs to attack the secondary targets—the Ennepe, Lister and Diemel dams.

The third wave of five aircraft (AJ–C, AJ–S, AJ–F, AJ–O and AJ–Y) would leave RAF Scampton more than two hours after the first two waves. They would follow the route of the first wave and act as mobile reserve to attack any of the primary targets that had not been breached or move onto the secondary targets. If all targets had been breached before they reached the Dutch coast, the reserve unit would be recalled.

All crews were warned not to stray from the planned routes because they were designed to avoid flak batteries, night fighter bases and hot–spots all the way from the Dutch coast to the dams and back. They would maintain low level during the whole flight there and back. They were also warned that under no circumstances should anyone return with a bomb intact. It was far too dangerous to attempt to land with an armed weapon. They were advised to release the bomb over preferably German land.

When it was his turn to address the aircrews, Wallis emphasized—as he had so many times before—the economic impact that taking out the Ruhr dams would have on the German war effort. He reviewed the prerequisites for effective delivery of the bomb—spinning at 500 rpm, dropped from 60 feet, at 230 mph. He went through it all again, perhaps with some relief that the day of the actual operation had finally come.

Finally, Wallis summed up 617 Squadron's role in the operation—'You gentlemen are really carrying out the third of three experiments,' Wallis told the airmen. 'We have tried it out on model dams, also one dam one—

fifth the size of the Möhne dam. I cannot guarantee it will come off, but I hope it will.' With those final words of background and encouragement Wallis sat down to gaze across the sea of young faces in front of him.

Up to that moment, most of the squadron's bomber crews had never known the word Upkeep, nor the notion of perhaps taking Germany out of the war. And they certainly hadn't known Barnes Wallis — the gentle, white-haired scientist who had designed this entire scheme and presented it to them that evening. It was just the latest of many revelations that would preoccupy them for at least the next 24 hours of their lives.

'It was like when you'd go in to write a school exam. Everybody's tense,' said Fred Sutherland, from Pilot Officer Les Knight's crew. For the front gunner, who had just turned 20, every piece of information at this briefing—just hours before they launched the operation—was entirely new. 'Everybody was pretty apprehensive about going over at low level and dropping this spinning bomb.'

As Joe McCarthy recalled, Wallis appeared to be a 'very mild and meek gentleman.' One of the officers present overheard Wallis comment softly in front of the assembled airmen, 'They must have thought it was Father Christmas speaking to them.'

Then Gibson stood up and for nearly an hour went back over the running order again. He emphasized the need for radio silence. He reviewed the known enemy defences and night-fighter locations, no doubt reminding many of the bomber crews present of the Ruhr's 'Happy Valley' moniker. On a more optimistic note, the weather remained ideal—skies would be clear, moon full, and winds negligible.

Gibson added that all the secrecy meant this raid against the dams could only be done once; a second attempt would have the enemy fully armed and ready. He finished by stating that no bomb was to be brought back to England. At about 1930 hours, some 90 minutes after it began, the squadron-wide briefing came to an end and the crews dispersed from the room and fanned out across the flight line into the early evening sunlight. It was two hours to take-off.

After the briefing the aircrews sat down to their evening meal in the mess. The mess staff recognized the significance of the fare they were asked to prepare—bacon and eggs, a luxury in Britain and an indication that an operation was about to be launched—but made no comments. Chiefy Powell had the mess staff also ensure that coffee, sandwiches and fruit were available for aircrews during the flight.

Then, most of the aircrews retired to their quarters to make their final preparations. Some wrote letters to their loved ones in case they did not make it back. Some made final meticulous inspections of their aircraft and weapon.

For all the married men at 617 Squadron there would be no time, or permission, for last words with wives or girlfriends. Any last correspondence, including wills and letters home, could be written but not posted or transmitted on this day. All outside calls were now prohibited, to family or anyone else. Station security even had RAF Scampton Women's Auxiliary Air Force (WAAF) Ruth Ive tapping all telephone calls that night, listening for indiscretions and disconnecting any such calls. The station was physically and electronically locked down.

Everything had been prepared and the crews were ready to make history...

## The Möhne Dam

The Möhne Reservoir is an artificial lake in North Rhine-Westphalia, some 45 km east of Dortmund, Germany. The lake is formed by the damming of two rivers, Möhne and Heve, and with its four basins stores as much as 135 million cubic metres of water.

*The Möhne Dam in the 1930's – note the substantial hydro-electricity building below the dam.*

In 1904 calculations about the future demand for water for people and industry in the growing Ruhr-area determined that the existing storage volume of 32.4 million m³ in dams of the Ruhr river system needed tripling. Thus, on November 28, 1904, the general assembly of the Ruhrtalsperreverein decided to construct additional dams.

During 1908 to 1913 they built the Möhnetalsperre at a cost of 23.5 million marks. When opened, the dam was the largest dam in Europe. 140 homesteads with 700 people had to move. It was built to help control floods, regulate water levels on the Ruhr river downstream, and generate hydropower. Today, the lake is also a tourist attraction.

The Möhne Dam was breached by RAF Lancaster Bombers

("The Dam busters") during Operation Chastise on the night of 16–17 May 1943, together with the Edersee dam in northern Hesse. Bouncing bombs had been constructed which were able to skip over the protective nets that hung in the water. A huge hole of 77 m by 22 m was blown into the dam.

*The Möhne Dam after the raid.*

The resulting flood wave killed at least 1,579 people, 1,026 of them foreign forced labourers held in camps down river. The small city of Neheim-Hüsten was particularly hard-hit with over 800 victims, among them at least 526 victims in a camp for Russian women held for forced labour.

Though the Organisation Todt quickly repaired the dams with 7,000 men from the construction of the Atlantic Wall, the impact of the raid on German industry in the Ruhr valley and indeed on the civil population was significant. According to Albert Speer, 'the power plant at the foot of the shattered dam looked as if it had been erased, along with its heavy turbines.' 'Industry was brought to a standstill', due to the 'electrical installations being soaked and muddied'.

The Möhne Dam was repaired by 23 September 1943, in time to collect water for needs the following summer, when the British failed to follow up with additional raids to hamper reconstruction.

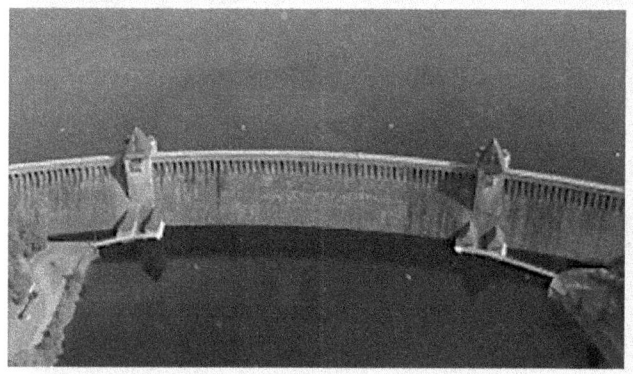

*The Möhne Dam today.*

## THE EDER DAM

The Eder Dam is a hydro-electric dam spanning the Eder river in northern Hesse, Germany. Constructed between 1908 and 1914, it lies near the small town of Waldeck at the northern edge of the Kellerwald.

*The Eder Dam before the war.*

At low water in late summers of dry years the remnants of three villages (Asel, Bringhausen, and Berich) and a bridge across the original river bed submerged when the lake was filled in 1914 can be seen. Descendants of those buried there go to visit the graves of their ancestors.

The dam was breached in the Second World War by 'bouncing bombs' dropped by British Lancaster bombers of No. 617 Squadron RAF as part of Operation Chastise. The early morning raid of 17 May 1943 created a massive 70 m wide and 22 m deep breach in the structure.

Water emptied at the rate of 8,000 cubic metres per second into the narrow valley below, producing a 6–8 m flood wave which roared as far as 30 km downstream. By the time it diminished in the widening floodplains of the lower Eder, into the Fulda and into the Weser, a total of about 160 cubic meters per hectare had flowed, wreaking widespread destruction and claiming the lives of some 70 people.

The dam was rebuilt within months by forced labour drawn from construction of the Atlantic Wall under command of Organisation Todt.

The Edersee lake today is the third largest reservoir in Germany. Its capacity of 199,300,000 cubic metres generates hydroelectric power and regulates water levels for shipping on the Weser river.

*The approach to the Eder Dam today with low water levels.*

*The Eder Dam today.*

## ROUTES

The map shows the approximate routes taken from RAF Scampton to the dams in the Ruhr Valley. The dashed line shows the route flown by the first and third waves. The dotted line shows the route taken by the second wave. The solid line shows the attack route on the primary targets (Möhne, Eder and Sorpe dams) taken by all waves. The two routes were chosen to suggest to enemy radar that these were minor attacks.

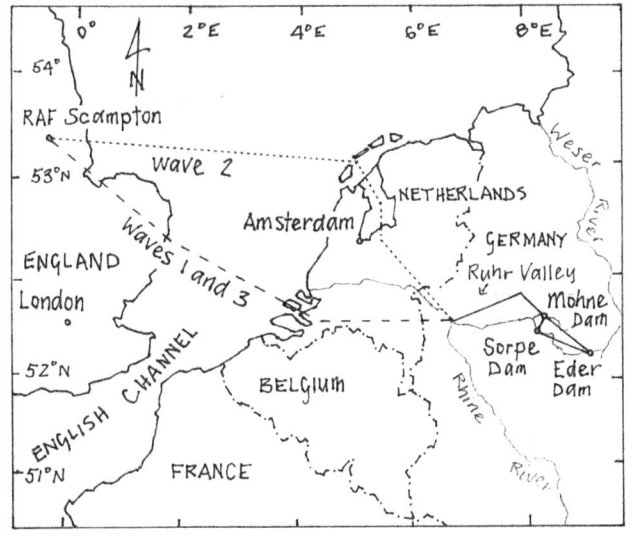

Locations of the dams and the route flown to attack the dams.

## Codewords

A code system was devised in order to relay messages back to Harris, Cochrane and Wallis at 5 Group operations room at RAF Grantham. A release of a bomb would be signalled by the prefix Goner followed by a number indicating what had happened after the release of the weapon.

| | |
|---|---|
| Goner 1 | - failed to explode. |
| Goner 2 | - overshot dam. |
| Goner 3 | - exploded more than 100 yards from dam. |
| Goner 4 | - exploded 100 yards from dam. |
| Goner 5 | - exploded 50 yards from dam. |
| Goner 6 | - exploded 5 yards from dam. |
| Goner 7 | - exploded on contact with dam. |
| Goner 8 | - no breach made. |
| Goner 9 | - small breach made in dam. |
| Goner 10 | - large breach made in dam. |

After Goner and the appropriate number, a letter was added to signify which dam had been attacked:

A - Möhne.

B - Eder.

C - Sorpe.

D - Lister.

E - Ennepe.

F - Diemel.

For example: a message Goner 79A would mean: weapon released at the Möhne Dam, exploded on contact with the dam and caused a small breach.

'NIGGER' – indicated a breach in the Möhne Dam and instructed the remaining aircraft with mines to attack the Eder Dam. 'DINGHY' – indicated a breach in the Eder Dam and instructed the remaining aircraft with mines to attack the Sorpe Dam.

# The Dams Raid

Just after 1300 hours on 16 May 1943, Bomber Command headquarters requested Fighter Command to launch intruder operations over the Continent in order to draw German night fighters away from 617 Squadron's attack routes through Holland to the Ruhr.

By 1400 hours, signals had been sent to USAAF, Fighter Command, Coastal Command and other bomber stations across the British Isles, warning of special ops that night.

Then, at 1515 hours 5 Group Headquarters contacted the Commanding Officer at RAF Scampton with the final op orders: 'Code name for 5 Group Operation Order B.976, is Chastise,' the telex message read. For the first time, Bomber Command had issued both the code number (the only title the operation had until then) and the code name. A cypher despatch — 'Executive Operation Chastise 16/5/43' — arrived next, meaning that 617 Squadron Lancasters were now under direct orders to take off that evening to complete the manoeuvres they had practised for seven weeks, this time as an attack on the major dams of the Ruhr River Valley.

The Squadron—wide briefing was held between 1800 and 1930 hours. By 2000 hours the No. 2 Hangar began to buzz again. Crews started to arrive, some on foot, others on bikes, to collect supplies and gear. In the crew room they gathered the usual equipment—jackets, boots, gloves, helmets, goggles and parachutes. Somebody pointed out that the squadron would be flying so low that even if a bomber survived a collision or crippling anti—aircraft fire, nobody would have sufficient time to clip on a parachute pack, scramble to a hatch, exit, survive the 200—mph slipstream, and then pop a chute. But training superseded all, this night; they gathered their parachutes anyway.

Around the crew room and out on the grass, the crews put on the faces they sensed would get them through the waiting. When he finished crossing off all the necessary items on his checklist, Adjutant Harry Humphries captured the demeanour of the aircrews in his squadron diary. 'Most of them wore expressions varying from 'don't care a damn' to the grim and determined,' he wrote. 'On the whole, I think it appeared rather reminiscent of a crusade.'

Just before 2030 hours Gibson pulled up in front of No. 2 Hangar in his car. He climbed out, followed by his entire crew. Humphries, in addition to noting the faces of the squadron aircrews, recorded that the squadron CO stood among his men looking 'fit and well and quite unperturbed.' As the crews milled next to the flight trucks, long-time comrades F/L John Hopgood and W/C Guy Gibson shared a last few minutes in the crew area. Gibson asked Humphries to ensure that there'd be plenty of beer for a party when they all got back. 'Hoppy,' Gibson called to his friend, 'tonight's the night. Tomorrow we will get drunk.'

Not wanting to reveal anything but confidence, Gibson soon announced that it was time to go. He joined others heading to the trucks that would taxi the aircrews in the first and second waves along the perimeter track around RAF Scampton airfield to their respective Lancasters. Among other things, Gibson thought about family; his Lancaster's registration letters, AJ—G, matched his father's initials, and 16 May was his father's birthday.

He thought about his canine pal, killed by accident the night before; Gibson had asked that the dog's body be buried on the grounds in front of his office at midnight, about the time the first wave would reach its target. And he wondered about his own fortunes; on this trip, as he always did, Gibson wore a Boy Scout badge on his right wrist as a lucky charm. Accompanying Gibson, Air Vice Marshal Cochrane and a photographer watched the aircrew prepare to board AJ—G and ensured that the Gibson was captured on film atop the entry ladder.

Along the perimeter, other crews and their gear were disgorged by the flight trucks to await the signal to board, start their Lancaster engines, and prepare for take-off. Outside Flight Lieutenant John Hopgood's, bomb aimer John Fraser and navigator Ken Earnshaw shared their feelings about the operation. The two Canadians had served together in 50 Squadron, and both men brought substantial hours of experience—Fraser and Earnshaw had flown 30 operations together. Still, as close as they were, Earnshaw occasionally caught Fraser off guard with his predilection for predicting the outcomes of combat operations. It happened again this night as the two men stood waiting for the signal to get underway. 'Ken, what do you think of tonight?' Fraser asked his buddy. 'Well, I think perhaps we might lose eight tonight ...' Earnshaw said, and before Fraser could react to the high casualty prediction, Earnshaw went further. 'And, you know, I think we might go ourselves.'

With half an hour to go before take-off, the crews of the first and second waves went to their aircraft. They went through the final checks, running the engines and then shutting them down ready for the signal to go. Just after 2100 hours, Hutchinson, Gibson's wireless operator fired a red light and the 52 Merlin engines on the 13 Lancasters of the first two waves roared as they prepared to depart.

The American pilot Joe McCarthy who should have been the first aircraft away had a last—minute problem when he discovered a coolant leak in the starboard (right) outer engine of AJ—Q during the warm up. There was no way AJ—Q would be able to fly, but luckily one spare aircraft had been flown in that afternoon and bombed up in anticipation of a problem somewhere. McCarthy and his crew switched to the spare AJ–T which had not been fitted with the spotlights or VHF radio because there had been no time. McCarthy reasoned that he did not need the spotlights because his target, the Sorpe Dam, did not require the height precision that the other gravity dams did.

Determined that he was not going to be left behind he and his crew unloaded everything they could from AJ–Q but in his rush, McCarthy pulled open his parachute. To add to his frustrations, when he reached AJ–T he found that the card giving the compass deviations was missing. Without it he would not be able to navigate correctly. Luckily the missing card was found and a replacement parachute was thrown to him as he climbed on board. McCarthy eventually left RAF Scampton 34 minutes late, after the first wave.

At 2128 hours a green Aldis light flashed from control and Barlow's AJ–E toiled along the grass runway with the rest of the second wave. They opened up their Merlin engines and the four Lancasters lifted off over the northern boundary fence and turned towards the North Sea. The first wave of nine aircraft soon followed in groups of three lead by Gibson, ten minutes apart.

After seeing the first two waves away at RAF Scampton, Wallis and Cochrane left to join Harris at the operations room at RAF Grantham.

The third wave which was acting as the reserve left RAF Scampton just after midnight, over two hours after the first two waves had departed. Before the last aircraft of the third wave had even lifted off, three crews of 21 men were already dead and two other Lancasters had aborted the mission.

In the two hours prior to the departure of the third wave, the first wave had flown out over the North Sea and the crews were testing their lamps and weapons. The winds were stronger than expected and it blew the first three aircraft of Gibson, Hopgood and Martin off course. They crossed the Dutch coast in the wrong position right over an area of anti—aircraft batteries. Luckily, they caught the gunners asleep and escaped without loss or damage.

They continued to occasionally drift off course as they crossed from Holland into Germany attracting the attention of flak and searchlights. During the flak attacks, Hopgood's aircraft was hit in the wing and he brushed over tree tops and under high tension cables as he struggled to keep the Lancaster in the air. During the journey to the Möhne, the gunners exchanged fire with defence position and in the intensity, they lost one another although they all arrived at the Möhne at roughly the same time.

Young, Maltby and Shannon were not far behind the leading group and were experiencing the same problems. They reached the Möhne eight minutes after Gibson relatively unscathed apart from a small hole in Shannon's Lancaster. Maudslay, Astell and Knight in the third flight had an uneventful fight until after they crossed the Rhine at Rees. At this point Astell crashed.

There are conflicting stories as to why Astell crashed. Bob Kellow, Knight's wireless operator said later that AJ—B was trailing behind when it was caught in a crossfire of light flak. German witnesses however, claim that Astell flew into an electricity pylon and crashed exploding into pieces as his bomb detonated. Astell and his entire crew were killed in the crash.

Astell was the fifth Lancaster to fall or turn back (all others being in the ill-fated second wave). By the time Maudslay and Knight arrived at the Möhne Dam the attack had already begun.

# The Möhne Dam Attack

It was a cloudless, moonlit night which gave Gibson an excellent view of the dam. The other Lancasters orbited the area some distance away and checked out the defences. Just as the reconnaissance had shown, an anti-torpedo net was floating in front of the dam but there were no search lights or balloons. There was light flak identified in each of the towers and along the wall as well as in the nearby countryside. There were an estimated 12 guns.

Gibson made a dummy run over the dam and then circled round and went in for his real attack. The German gunners were now ready for him and were probably amazed to see the Lancaster turning on its lamps as it approached the dam. With the bomb now being revolved backwards at 500 rpm in the belly of the aircraft, Taerum—the navigator—talked Gibson down to the required 60 feet. At 0028 hours Spafford the bomb aimer pressed the release button and the bomb was away for real.

Just as it did in testing, the bomb bounced across the lake three times before sinking down the centre of the dam. Moments later a huge column of water was thrown up as it exploded, momentarily obscuring Trevor-Roper's view of the dam from the rear turret of AJ–G.

Everyone thought that the huge dam had given way but as the water and spray subsided, they realised that the dam wall was still intact. Hutchinson,

Gibson's wireless operator sent back the Morse message 'Goner 68A'—weapon released against Möhne dam, exploded five yards from dam with no breach.

Gibson gave the water a few minutes to settle before calling Hopgood in AJ—M to attack just as Maudslay and Knight were arriving. After being slightly damaged on the journey to the dam, AJ—M was punished further from the German gunners. The other planes saw shells hitting the port (left) engines and going into the starboard (right) wing. With the battering and flames now coming from the engines, Fraser released the bomb late. The bomb bounced straight over the dam and hit the power station on the other side. AJ—M now crippled, limped over the dam in a ball of flames after a fuel tank had no doubt been hit.

Knowing he was doomed Hopgood tried to gain some height to allow his crew to bail out. Tony Burcher in the rear gun turret managed to escape by cranking the turret around by hand after the hydraulics had been knocked out due to the loss of the port (left) engine. He crawled into the rear fuselage where his parachute was stored. As he strapped it on, he saw John Minchin the wireless operator crawling down towards him seriously wounded. Burcher bravely pushed his fellow crew mate out of the rear door, pulling the rip cord of the parachute as he fell. Realising they were now too low, Burcher opened his parachute while inside the plane throwing it out of the door and letting it pull him clear of the doomed Lancaster. As he left the plane the wing sheared off and he passed out.

AJ—M plunged into the ground in ball of flames three miles from the dam. Hopgood, Earnshaw, Brennan and Gregory were killed instantly. Burcher came around on the ground but had sustained serious back injuries from either hitting the tail of the plane or landing heavily. Minchin's parachute had failed to open in time and he was killed. Fraser the bomb aimer had also managed to escape from the Lancaster in the same way Burcher had by allowing the parachute to pull him clear of the plane through the front escape hatch. Both Burcher and Fraser were taken prisoner.

At the moment AJ—M hit the ground the self—destruct fuse on their bomb had detonated it destroying the power station on the dry side of the dam.

There was no time to mourn for the remaining crews, they still had a job to do. Gibson called in Micky Martin—the low flying expert—in AJ—P. In an attempt to reduce the amount of flak Martin would get and try to prevent a repeat of the previous disaster with Hopgood and AJ—M, Gibson flew in just ahead of him drawing the flak away. Martin made a perfect run and released his bomb at the correct point but it veered off to the left and exploded short of the dam with no effect. As Martin flew over the dam at the end of his attack some flak shells hit his starboard (right) wing and punctured a fuel tank. Luckily for AJ—P the tank was empty. 'Goner 58A' was relayed back to Grantham—weapon released at Möhne dam, exploded 50 yards from dam with no breach made.

Dinghy Young was the fourth plane to attack in AJ—A. Despite the damage sustained during his attack Martin in AJ—P joined Gibson in trying to distract the gunners on the dam while Young made his attack. This time Gibson turned on his navigation lights to draw the German's attention while MacCausland, Young's bomb aimer released the bomb. Their attack was spot on, the bomb bounced three times, sank down next to the dam wall and exploded throwing up another huge column of water. Again, the crews must have thought that the dam had gone but when the water subsided, they found to their dismay that the structure was still intact. Goner 78A was sent to Grantham—weapon released at Möhne Dam, exploded on contact with the dam but no breach made.

There was an air of gloom descending over RAF Grantham and frustration with the crews at the dam. There had been one perfect hit and another close one yet still the dam stood firm. After all, Wallis believed only one bomb would be required to punch a hole through the dam.

Undeterred Gibson radioed Dave Maltby into attack in AJ—J. With Gibson and Martin drawing fire on either side Maltby went in for his run.

Just as he was about to release his bomb, Maltby noticed that the crown of the dam had begun to crumble and the centre had opened up slightly. Young's bomb had worked! AJ—J continued the attack never the less and with another perfect hit the dam gave way with spectacular effect. For some reason another 'no breach' message was sent back to Grantham after the attack, probably because the scene was still obscured. Gibson however soon saw a river of water rushing through the dam wall sweeping it away down the valley. He called off Shannon who was about to make his attack and ordered his wireless operator Hutchinson to send 'NIGGER' back to Grantham— the codeword for the successful breach of the Möhne Dam.

The scene at RAF Grantham operations room was one of relief and joy as the codeword 'NIGGER' came through. Harris who had not so long before called Wallis' idea 'Tripe of the wildest description' now shook his hand and exclaimed 'I didn't believe a word you said when you came to see me, but now you could sell me a pink elephant'. While RAF Grantham operations room celebrated the success, the remaining aircraft in the first wave set off for the next target—the Eder Dam.

*The Operations Room at RAF Grantham.*

# THE EDER DAM ATTACK

After seeing the effects of their work circling the Möhne, Gibson sent Martin in AJ—P and Maltby in AJ—J home and ordered the remaining Lancasters to follow him to the Eder Dam.

Shannon in AJ—L, Maudslay in AJ—Z and Knight in AJ—N with bombs followed Gibson and Young as they turned south—east for the short flight to the Eder. Despite releasing his bomb at the Möhne Dam, Young went on to the Eder because he would take over as leader should anything happen to Gibson.

The Eder Dam was difficult to locate with the similarly looking wooded landscape with valleys and an early morning mist rising. Having located the target, the Lancasters began to circle to evaluate the task ahead. A bonus for the crews was that there were no flak gunners over the Eder defending the dam; the dam was undefended.

However, the terrain was not going to make it an easy task by any stretch of the imagination. In fact, it was probably more difficult to get a good run. Due to the shape of the valley, the attackers would have to approach over Waldeck Castle sat on top of a 1,000 feet peak, then dive down to the lake and swing sharply left, hop over a spit of land and quickly drop to 60 feet for the attack. As soon as they released their bomb, they would then have to pull up steeply in order to avoid the high ground on the other side of the dam. In actual fact they would have about five seconds to line up the plane at 60 feet and release the weapon before it was too late.

At 0120 hours Shannon in AJ—L was sent in to attack the Eder. After three attempts at lining up correctly he could not achieve the right angle or height. Gibson told him to take a breather and circle round while Maudslay in AJ—Z was sent in for an attempt. Maudslay had similar challenges and after two attempts Shannon was told to try again.

Shannon made two more runs but didn't manage to release his bomb. On this third attempt he lined up well and released his bomb. It exploded right against the dam wall sending a column of water 1,000 feet into the air. But the dam held. The codeword 'Goner 78B' was send to Grantham—weapon released at Eder Dam, exploded against dam but no breach.

Maudslay was sent back in by Gibson for another attempt. He was still struggling to get the correct height when his bomb aimer Mike Fuller, perhaps not wanting to delay the attack released the bomb too late with fateful effect. It hit the parapet of the dam and exploded in a burst of light which by all accounts illuminated the whole valley like daylight.

Although Maudslay had just cleared the dam when the bomb exploded, the blast almost certainly caused damage to AJ—Z. Gibson radioed him over the R/T and asked if he was OK. Maudslay was heard to faintly reply 'I think so'. It was the last they would hear from Maudslay and the crew of AJ—Z.

Gibson then called up Astell in AJ—B who he had not seen since leaving RAF Scampton, but Astell had crashed into a pylon on the flight to the targets. With only one bomb remaining from the nine aircraft assigned the Möhne and Eder dams and the Eder still standing it was all down to Les Knight and AJ—N.

Knight made one dummy run to get a feel for the approach. As he flew over the dam Knight opened the throttles of the Lancaster and stood it on its tail as he recovered the plane on the far side of the dam. The climb after the attack run was hair raising. Johnson said later that it 'required the full attention of the pilot and engineer to lay on emergency power from the engines and a climbing attitude not approved in any flying manuals and a period of nail biting from the rest of us not least me who was getting too close a view of the approaching terra—firma from my position in the bomb—aimer's compartment.' Harry O'Brien, the rear gunner in AJ—N, later recorded he 'never thought they would get over the mountain' on the other side of the dam.

On his second run, with a bright moon on the starboard beam, Knight lined AJ—N up perfectly and Ed Johnson the bomb aimer released the bomb spot on. The bomb skipped over the lake three times and hit the wall not far from the centre.

*A painting of the moment when AJ—N breached the Eder Dam.*

Behind them, the bomb exploded throwing up a huge column of water. To the delight of all the crews the bomb punched a hole right through the middle of the dam and then the top fell away with a gigantic torrent of water bursting through. Knight later recalled that the explosion caused a 'large breach in [the] wall of [the] dam almost 30 ft below top of [the] dam, initially leaving top of [the] dam intact.'

Bob Kellow had his head up in the astrodome, looking backwards. It seemed, he said, 'as if some huge fist had been jabbed at the wall, a large almost round black hole appeared and water gushed as from a large hose.'

Fred Sutherland recalled in a later interview 'We were all afraid of the hill. We had to drop the bomb at the right distance and the right height, and then to make it [Les] had to push the throttles right through the gate, which is not supposed to be done…I didn't see anything when the bomb went off because I was in the nose, but I heard the rear gunner saying: "it's gone, it's gone".'

Gibson then ordered the crews of the remaining Lancasters to make their way home by the pre-arranged escape routes. For most of them it was an uneventful journey home with the occasional searchlight or burst of flak.

AJ–N headed for home via the Möhne Dam, where they noticed how much the water level had already dropped. The trip back was relatively trouble-free—they avoided some flak bursts near Borken, and Fred Sutherland was able to shoot up a stationary train in a small town.

They were very lucky, however, not to have fallen at the final hurdle in an incident which only O'Brien noticed: '… at the Dutch coast the terrain rose under us, Les pulled up, over and down. On the sea side of this rise was a large cement block many feet high. This block passed under our tail not three feet lower. As the rear gunner I was the only one to see it.'

Two of the crews were not so lucky however. Maudslay in AJ–Z who had been heard limping away from the Eder after being caught in the blast of the bomb had been believed to have crashed near the dam. It later emerged however that the damaged AJ–Z had flown on for another 45 minutes. 140 miles away from the Eder and half way to the coast Maudsaly was caught in light flak and at 0236 hours crashed at Emmerich just inside the German border. Maudslay and all of his crew perished in the crash.

Twenty minutes later Young in AJ–A also ran into trouble. Young was given the nick name 'Dinghy' after twice bailing out into the sea and being saved by his dinghy. Young had flown to the Eder unarmed after weakening the Möhne with his bomb. He was acting as the deputy leader, ready to take over if anything happened to Gibson. After the Eder was breached Young turned for home and as they neared the Dutch coast they came under fire. Young's wingmen Maltby and Shannon had seen that he had been flying too high all night—maybe having trouble finding navigational landmarks. This may have been his undoing when he flew over the heavily defended coastal area north of Ijmuiden and was shot down by the flak batteries stationed there. AJ—A did clear the coast but plunged into the North Sea with the loss of all crew.

# The Sorpe Dam Attack

The attack of the Sorpe by the second wave can only be described as cursed from the start, from the moment Joe McCarthy's AJ—Q sprang a coolant leak during the warm up and he was forced to switch to the reserve AJ—T. After being delayed by 34 minutes by the problem, McCarthy was determined to make up the time and catch the other four Lancasters in the wave. McCarthy soon went from being late to being the only plane left in the second wave.

The second wave were unable to fly together as the tactics of the raid called for them to depart alone and take a more northerly route over the North Sea. Their attack technique was also going to be different as they were attacking an earthen dam. They would have to explode the bomb near the crest of the dam rather than the base where the earth would absorb the shock waves. The bomb would also not be rotated unlike attacking the other dams. It would be more like a conventional dive—bombing manoeuvre. In order to maximise the attack area, the crews of the second wave had practiced attacking the dam along the length rather than perpendicular (at right angles) to it. With the dam located in a steep valley this was going to require exceptional flying skills similar to those employed at the Eder. They would have to drop down over one valley wall and release the weapon before hauling the Lancaster back up the other valley side.

While McCarthy was busy switching to the spare aircraft, the lead of the wave fell to Bob Barlow in AJ—E. He departed RAF Scampton at 2128 hours and was never seen again.

Vernon Byers in AJ—K was the third aircraft to depart RAF Scampton, again nothing more was heard from him. It appears that over the Dutch coast he strayed off course over the island of Texel which was a notorious flak hot spot. He climbed higher to try to get a fix on his position but the gunners shot him down into the Zuider Zee at 2257 hours. AJ—K was the first aircraft lost in the operation with the loss of all crew.

At about the same time as AJ—K was coming under fire, Les Munro in AJ—W ran into similar trouble. As he roared over the island of Vlieland just north of Texel, AJ—W also became an easy target for land—based gunners and a flak ship. It was a miracle that no one was killed but the electrics in the plane were knocked out killing communication with other aircraft and his own crew. In an impossible position to attack the dams having no communication he thought he would only be a menace to other aircraft and reluctantly turned home. He nearly had another disaster when he returned to RAF Scampton, he flew straight in cutting up another Lancaster also attempting to land after aborting its mission. A collision would have been disastrous as Munro had, against orders, returned with his bomb armed and dangerous.

The other Lancaster of the second wave was Geoff Rice in AJ—H. He too had been forced to turn back after another mishap. After crossing Vlieland where the crew were witness to Byers' ending in AJ—K, Rice flew dangerously low to the water in an attempt to outfox the gunners. He overdid it somewhat and with the altimeter on zero he clipped the sea. The bomb he was carrying was ripped from the belly of the Lancaster and vanished into the water. The bomb also damaged the tail wheel as it was torn off. The rear of the Lancaster was filled with sea water and the badly shaken Rice struggled to bring AJ—H up with water streaming out of the empty bomb bay. They were all lucky to survive but without a bomb their mission was useless so they too turned home.

Rice had to make an emergency landing when he returned to RAF Scampton with a damaged rear landing wheel and having to lower the undercarriage manually after a hydraulics failure. Fearing a rough landing, the crew took up the crash positions as Rice gingerly put the Lancaster down safely but with one final scare as the radio less Munro, unable to warn anyone he was landing shot in below him.

As Rice and Munro were heading home another Lancaster from their wave was falling. Barlow who had departed first had just crossed the German border at 2350 hours when close to Rees the aircraft came into trouble. The British report that AJ—E had come under fire and was hit by flak. The Germans however reported that the aircraft simply flew into an electricity pylon. However, AJ—E met its end, it took Barlow and his whole crew with it. There were no survivors.

Although AJ—E disintegrated, the bomb it was carrying did not break free of the aircraft with the result that the self—destruct mechanism did not arm. Given that the impact was not severe enough to explode the bomb also suggests that Barlow had tried to land the doomed AJ—E in the few seconds he had left. The Germans now had in their possession the RAF's top—secret weapon intact and it would not take them long to figure out how it worked.

By now four of the five Lancasters from the second wave were either shot down or heading home due to problems. McCarthy who had departed 34 minutes late in the spare AJ—T was all that was left of the second wave. He kept as low as possible dodging flak and searchlights which had been alerted by the aircraft that had gone before. McCarthy also noticed that at times he flew below packs of night fighters flying 1,000 feet above him.

He finally reached the Sorpe Dam at 0015 hours with more problems ahead. The approach to the dam looked even worse that on the models and photo reconnaissance. McCarthy had to line up his approach over a church steeple in the village of Langscheid on the crest overlooking the dam before swooping down into the valley. With only a matter of seconds before the aircraft had to pull up to avoid flying into the other valley side at the far end of the dam, bomb aimer George Johnston had no time at all to talk his pilot onto the correct line and height for release. It was even more difficult as this aircraft was not fitted with the spot lamps.

They tried to line up correctly nine times and every time Johnston was not happy. Finally, on the tenth run Johnston released the bomb. As the Lancaster climbed the valley on the far side, the bomb exploded and a plume of water followed. They turned back to assess the damage and saw that a section of the crest had been blow away. To their disappointment, the Sorpe Dam was still standing.

On the way home McCarthy strayed over some heavy flak positions especially at Hamm. His evening had one more twist when he was forced to land at RAF Scampton on a bullet burst tyre. For unknown reasons McCarthy failed to send the codeword 'Goner 79C'—Weapon release at the Sorpe Dam, exploded on contact with the dam and a small breach made until he was 20 minutes from home. As a result, controllers at RAF Grantham were not aware that the Sorpe Dam, which was the second most important target after the Möhne Dam, was still standing. So, despite this primary target remaining, the third wave continued their attack onto the three secondary targets; the Diemel Dam, the Lister Dam and the Ennepe Dam.

But one of the five Lancasters in the third wave—Freddie Brown in AJ—F—was able to be redirected to the Sorpe Dam. He attacked it at 0314 hours but was not able to breach it. There were no other aircraft available and so the third of the primary targets had survived the night's attacks.

Bill Townsend in AJ—O was the only aircraft to attack any of the three secondary targets. They attacked the Ennepe Dam at 0341 hours but it was not breached.

# Returning to RAF Scampton

Maltby was the first of 11 Lancasters that began landing back at RAF Scampton at 0311 hours on 17 May. Gibson landed at 0415 hours and Les Knight five minutes later at 0420 hours. On the return journey Gibson had seen an aircraft falling in flames over Hamm. Little did he know that it was Warner Ottley in AJ—C of the third reserve wave still on his way to the Lister Dam.

Townsend's AJ—O landed at 0615 hours, the last to return because one of its engines had been shut down after passing the Dutch coast. With Townsend's return, the full extent of the losses—eight aircraft and 56 crewmen—became apparent.

Air Chief Marshal Harris was among those who came out to greet the last crew to land. Barnes Wallis was distraught when he learned the human cost, 'What have I done that I have caused these young men to be sent to their deaths?' Gibson told him that no crew ever went on a mission without the thought they might not return.

Ed Johnson, who knew Wallis in his later years, related that the designer never forgave himself for the losses, and dedicated his development of the 'Tallboy' and 'Grand Slam' bombs to the hope they would end the death and destruction of the war sooner.

After debriefing many of the officers' retired to the Officers' Mess to celebrate their success. Harold Hobday took part in the celebrations with a fair degree of gusto. He was photographed outside the Officers' Mess about breakfast time but fell asleep in an armchair sometime later and regained consciousness after lunch time.

*The 'morning after the raid' photograph taken outside the RAF Scampton Officers' Mess. Les is second from the right in the front row.*

# Accolades

News of the successful raid spread quickly and the RAF was quick to capitalise on the story. *The Times* reported on 18 May 1943 that:

> 'The R.A.F. struck what the Secretary of State for Air has described as 'a trenchant blow for victory' when early on Monday morning Lancaster bombers breached three dams which serve the Ruhr industries. The attacks were made by specially selected crews and the bombers came as low as 10ft. to plant their mines on the lips of the dams. In one a breach of 100 yards was caused and the released waters rushed down into the valleys, sweeping all before them. Reconnaissance flights yesterday showed that vast damage had been done by flood waters sweeping down the Ruhr valley.'

*The front page of the* Daily Telegraph *on 18 May 1943.*

*Les speaking with King George VI on 27 May 1943 during the King and Queen's visit to RAF Scampton.*

On 27 May 1943 King George VI and the Queens visited RAF Scampton to meet the crews and receive a brief on the raid. During the visit the King approved a new badge depicting the breached Möhne Dam and the motto *Après moi le déluge* (After me, the flood).

A number of awards were announced. Guy Gibson was awarded the Victoria Cross—making him the most highly decorated member of the RAF.

*The crew of AJ—N circa July 1943: Back Row (L to R) Harold Hobday, Ed Johnson, Fred Sutherland, Bob Kellow, Ray Grayson. Front Row (L to R) Les Knight, Henry O'Brien.*

Les was awarded the Distinguished Service Order, and navigator Harold Hobday and bomb aimer Edward Johnson both awarded the Distinguished Flying Cross. Les' citation read:

> On the night of 16 May, 1943, a force of Lancaster Bombers was detailed to attack the Möhne, Eder and Sorpe dams in Germany. The operation was one of great difficulty and hazard, demanding a high-degree of skill and courage and close cooperation between the crews of the aircraft engaged. Nevertheless, a telling blow was struck at the enemy by the successful breaching of the Möhne and Eder Dams. This outstanding success reflects the greatest credit on the efforts of the personnel who participated in the operation in various capacities as members of aircraft crews.

Les was embarrassed that the whole crew had not been rewarded. Fred Sutherland later recalled: 'He felt badly that half the crew got decorated and the other half didn't. He said you know I'm wearing the DSO for all you guys, you all did something for it.'

On 22 June 1943 Les, Sidney Hobart and Ed Johnson along with the others from the Dams Raid that had been awarded decorations were invested by the Queen at Buckingham Palace.

*Sidney Hobart, Les Knight and Ed Johnson outside Buckingham Palace on the day of their investiture.*

*Above:* Some of the Australians on the dams raid. L. to R. : Bob Hay, Lance Howard, David Shannon, Jack Leggo, Spam Spafford, Micky Martin, Les Knight and Bob Kellow.

A publicity photograph of the Australians who survived the Dams Raid.

617 Squadron Aircrew after the Dams Raid.

Portrait of Les Knight by Cuthbert Orde.

After the investitures all the crews went on to the Hungaria Restaurant to a dinner arranged by AV Roe, the builders of the Lancaster; which Les notably skipped.

In July 1943, several of the Dams Raid crews—including AJ—N—were required to pose for RAF publicity photographs. Les even had his portrait done by Cuthbert Orde on 27 June 1943. Orde was a former Royal Flying Corps pilot in the First World War who completed hundreds of portraits of pilots during the Second World War.

# Impact

The two direct hits on the Möhne Dam resulted in a breach around 76 metres wide and 89 metres deep. The destroyed dam poured around 330 million tons of water into the western Ruhr region. A torrent of water around 10 metres high and travelling at around 24 km/h swept through the valleys of the Möhne and Ruhr Rivers.

A few mines were flooded; 11 small factories and 92 houses were destroyed and 114 factories and 971 houses were damaged. The floods washed away about 25 roads, railways and bridges as the flood waters spread for around 80 km from the source. Estimates show that before 15 May 1943 water production on the Ruhr was 1 million tonnes; this dropped to a quarter of that level after the raid.

The Eder River drains towards the east into the Fulda River which runs into the Weser River then on to the North Sea. The main purpose of the Eder Dam was then, as it is now, to act as a reservoir to keep the Weser and the Mittelland Canals navigable during the summer months. The wave from the breach in the Eder Dam was not strong enough to result in significant damage by the time it hit Kassel (approximately 35 km downstream).

The greatest impact on the Ruhr armaments production was the loss of hydroelectric power. Two power stations (producing 5,100 kilowatts) associated with the dam were destroyed and seven others were damaged. This resulted in a loss of electrical power in the factories and many households in the region for two weeks. In May 1943 coal production dropped by 400,000 tons which German sources attribute to the effects of the raid.

According to an article by German historian Ralf Blank, at least 1,650 people were killed: around 70 of these were in the Eder Valley, and at least 1,579 bodies were found along the Möhne and Ruhr rivers, with hundreds missing. 1,026 of the bodies found downriver of the Möhne Dam were foreign prisoners of war and forced labourers in different camps, mainly from the Soviet Union. Worst hit was the city of Neheim at the confluence of the Möhne and Ruhr Rivers, where over 800 people perished, among them at least 493 female forced labourers from the Soviet Union.

After the operation Barnes Wallis wrote, 'I feel a blow has been struck at Germany from which she cannot recover for several years', but on closer inspection, Operation Chastise did not have the military effect that was at the time believed. By 27 June 1943, full water output was restored, thanks to an emergency pumping scheme inaugurated the previous year, and the electricity grid was again producing power at full capacity.

The raid proved to be costly in lives (more than half the lives lost belonging to Allied POWs and forced-labourers), but was no more than a minor inconvenience to the Ruhr's industrial output. The value of the bombing can perhaps at best be seen as a boost to British morale. Critics believed the raid was oversold, its achievements exaggerated and other Bomber Command raids unfairly ignored.

In his book *Inside the Third Reich*, Albert Speer acknowledged the attempt: 'That night, employing just a few bombers, the British came close to a success which would have been greater than anything they had achieved hitherto with a commitment of thousands of bombers.' He also expressed puzzlement at the raids: the disruption of temporarily having to shift 7,000 construction workers to the Möhne and Eder repairs was offset by the failure of the Allies to follow up with additional (conventional) raids during the dams' reconstruction, and that represented a major lost opportunity. Barnes Wallis was also of this view; he revealed his deep frustration that Bomber Command never sent a high—level bombing force to hit the Möhne Dam while repairs were being carried out. He argued that extreme precision would have been unnecessary and that even a few hits by conventional High Explosive bombs would have prevented the rapid repair of the dam which was undertaken by the Germans.

The Dams Raid was, like many British air raids, undertaken with a view to the need to keep drawing German defensive effort back into Germany and away from actual and potential theatres of ground war, a policy which culminated in the Berlin raids of the winter of 1943–1944. In May 1943 this meant keeping the Luftwaffe aircraft and anti-aircraft defences away from the Soviet Union; in early 1944, it meant clearing the way for the aerial side of the forthcoming 'Operation Overlord'. The considerable amount of labour and strategic resources committed to repairing the dams, factories, mines and railways could not be used in other ways, on the construction of the Atlantic Wall, for example. The pictures of the broken dams proved to be a propaganda and morale boost to the Allies, especially to the British, still suffering from the German bombing of the Baedeker Blitz that had peaked roughly a year earlier.

Another effect of the Dam Raid was that Barnes Wallis's ideas on earthquake bombing, which had previously been rejected, came to be accepted by 'Bomber' Harris. Prior to this raid, bombing had used the tactic of area bombardment with many bombs, in the hope that one would hit the target.

Work on the earthquake bombs resulted in the 'Tallboy' and 'Grand Slam' weapons, which caused damage to German infrastructure in the later stages of the war. They rendered the V-2 rocket launch complex at Calais unusable, buried the V-3 guns, and destroyed bridges and other fortified installations, such as the 'Grand Slam' attack on the railway viaduct at Bielefeld. Most notable successes were the partial collapse of 6.1 m, reinforced concrete roofs of U—boat pens at Brest, and the sinking of the battleship *Tirpitz*.

*The Möhne Dam after the raid.*

*The Eder Dam after the raid.*

IMPACT 107

*The remains of the hydroelectric station at the Eder Dam.*

*The Eder Dam being repaired.*

*Damage downstream from the Möhne Dam.*

# Guy Gibson–Part 2

On 18 May, there was a parade where Cochrane and Gibson made speeches to the squadron members. He then released the air crews from duty on seven days leave and half the ground crew on three days leave. Gibson went on weekend leave to Penarth. On the Sunday he received a call from Harris to inform him he had been awarded the Victoria Cross (VC). His response was subdued as he felt responsible for those he had recruited and who had not returned, particularly Hopgood. He was reported as saying: 'It all seems so unfair'.

On 22 June, Gibson and all the other newly decorated members of 617 Squadron attended an investiture at Buckingham Palace. It was performed by the Queen as the King was in North Africa. She presented Gibson with his VC and the Bar to his DSO first, and in the process, he became the most highly decorated serviceman in the country. After the investitures all the crews went on to the Hungaria Restaurant to a dinner arranged by AV Roe, the builders of the Lancaster. Gibson was presented with a silver model of a Lancaster by the company's chairman, Thomas Sopwith. Also, at the dinner were Roy Chadwick, the designer of the Lancaster, and Wallis.

Harris made arrangements to ensure Gibson was rested from operations and on 24 July he and his wife were invited to lunch at Chequers as guests of the Prime Minister, Winston Churchill. Here Gibson was shown a film smuggled out of Germany about the concentration camps. On 2 August, Gibson made his last fight with 617 Squadron. He flew with his regular crew and his successor, Wing Commander George Holden, to Eyebrook Reservoir to familiarise him with the technique to release the bouncing bomb.

**Tour of Canada and US**

On 3 August Gibson travelled to London to join the party accompanying the Prime Minister to the Quadrant Conference in Quebec, Canada. Around midnight they were taken by a special train to Faslane where they boarded the *Queen Mary*. The party included some of the most senior military figures such as Lord Louis Mountbatten, Chief of Combined Operations and Air Chief Marshal Sir Charles Portal, Chief of the Air Staff. Gibson was therefore an outsider, like fellow passenger, Brigadier Orde Wingate, the Chindits leader. However, unlike Wingate, he seems to have found it easier to enter into shipboard life. Mary Churchill, who was travelling as her father's aide-de-camp, found Gibson 'had all the aura of a hero' and also 'very agreeable and debonair to talk to'. On the last evening of the voyage, on 8 August, after dinner, Gibson was invited to address the whole party on the Dams Raid.

On 9 August they arrived in Halifax, Nova Scotia, and were transferred by special trains to Quebec. A certain amount of disinformation circulated around their arrival including how Gibson had acted as the pilot on the aircraft that had flown Churchill across the Atlantic. They arrived at a time of significant tension between the British and Canadian governments. The Canadians were unhappy with the relative lack of credit being granted to the Royal Canadian Air Force's (RCAF) contribution to the war effort. On 11 August, Gibson attended a select luncheon with the Prime Minister, where he was introduced to the Canadian Prime Minister, Mackenzie King. He spent the rest of the afternoon at an RCAF recruiting centre.

On 12 August Gibson attended a Press Conference arranged in his honour. It was hosted by Hon. C.G. Power, the Canadian air minister and attended by hundreds of journalists from around the world. Gibson responded to questions about the Dams Raid and revealed the Prime Minister called him 'Dam-buster'. Reports of the conference were enthusiastic. He attended engagements in the Quebec area. On 17 August, President Franklin D. Roosevelt arrived at the Conference. Churchill arranged for Gibson to meet the President at a private meeting.

Gibson left Quebec on 20 August to start the Canadian leg of his tour. It was a punishing schedule and included civic receptions and speeches, motorcades, interviews, broadcasts, along with travel. He went to Montreal, Ottawa, Toronto and London (Ontario). As Churchill was in New York, on 5 September, Gibson was diverted there to make a radio broadcast which was heard on station WJZ New York. In Winnipeg he met the family of Harvey Glinz, who had been killed on the Dams Raid when flying as the front-gunner in Barlow's crew. He then went on to training bases at Carberry, Rivers, Dafoe, Moose Jaw and Moss Bank.

On 11 September he arrived in Calgary. Here he met the mother of the navigator in his own Dams Raid crew, Mrs Taerum. He spent time with her at home the following day. He also met Leading Aircraftman Robert Young, the younger brother of Squadron Leader Young, also killed on the Dams raid. He continued on to Vancouver and Victoria before returning to Montreal and Ottawa in the east. He had a week's rest at the Seignory Club in Ottawa from 28 September to 3 October.

On 4 October he began the United States leg of his tour in Washington, D.C. He attended a major Press Conference at the offices of the British Information Service in New York on 7 October. This was at a time when the first American airmen were coming home 'tour expired' after 25 operations. During questions one young lady asked, 'Wing Commander Gibson, how many operations have you been on over Germany?' He replied, 'One hundred and seventy-four.' There was a stunned silence.

On 19 October, Gibson was invested with the Commander's Insignia to the Legion of Merit by General Henry H. Arnold at Bolling Field near Washington D.C. The decision to award him with the Legion of Merit was taken quickly. It was also exceptional. To avoid duplication American awards were, in general, not accepted for actions which had already earned a British medal. This allowed them to be restricted to cases where Americans wished to express particular gratitude. For example, they were often given for the air/sea rescue of American personnel. It was announced formally in Britain in December 1943.

Gibson continued on to Chicago, Minneapolis and then to Los Angeles. He stayed with the film director Howard Hawks. Most his time was spent in private, his reward for his gruelling tour. However, it is possible he might have been giving technical advice on a proposed film of the Dams Raid. Hawks had commissioned Roald Dahl to write a script for the film and had started to build models of the dams and Lancaster bombers. He was encouraged by Bomber Command's PR Department. However, when Wallis was shown the script, he thought it absurd, and the project was scrapped.

Gibson returned to Montreal and flew back to Britain in a B-24 Liberator being ferried across the Atlantic. He landed at Prestwick on 1 December and was met by a representative from the Ministry of Intelligence. On his return he was exhausted, but still went straight from home in London to Scampton to visit 617 Squadron. When he arrived, he was informed that they had moved to Coningsby, and was driven over. He visited HQ 5 Group in an attempt to obtain an operational posting, but was declared non-operational sick and sent on a month's rest leave. At the end of his leave he was due to be posted to the Directorate of Accidents with the order to write a book. During this time, he was hospitalised with Vincent's Angina on 17 December.

The view emerged that as a result of the tour he had acquired and retained an increased sense of his own importance. In July 1944 Harris wrote to Cochrane to comment that the Americans had 'spoiled young Gibson'. Therefore, this route was not pursued again later in the war with other highly decorated airmen, such as Leonard Cheshire.

### Directorate for the Prevention of Accidents: writing Enemy Coast Ahead

In January 1944 Gibson was posted to the Directorate for the Prevention of Accidents. He appears to have been under orders to write a book. This posting was effectively a cover to give him the time and access to the resources he needed to complete it. It is possible either the Ministry of Intelligence or the RAF's publicity department wanted him to complete a book in order to counter the increasing criticisms of the Strategic Air Offensive.

Gibson was located in a small back room and had access to a dictaphone and typist. He did not seem to take well to his assignment initially. When Heveron travelled from RAF Scampton to deliver some information about 617 Squadron, he found him depressed and with long hair.

However, Gibson did seem to become increasingly enthusiastic about writing. His wife remembered him writing at home during weekends while he was at Staff College during March–May 1944. The typescript survives of a draft Gibson submitted in summer 1944. His wife donated it to the RAF Museum at Hendon. The writing style confirms that Gibson wrote most of the book, because it includes his characteristic style of comments and humour, plus in places it is simply bad. Therefore, the book was not ghost-written, as some have suggested. The typescript includes corrections in his own hand. These may suggest he had the help of a professional editor while he was writing. The text was passed by the censors who edited out some of his more intolerant political and social views as well as operational information. He completed his final draft in September 1944.

Gibson attended a staff course at the RAF Staff College at Bulstrode Park near Gerrards Cross from the end of March to May 1944. He then went on leave. During the last week he got very restless as he learnt about the D-Day landings. He feared the war would end before he could get back into the action. On his return he appealed straight to Harris. Four days later he was appointed as a staff officer at No. 55 Base, RAF East Kirkby to understudy to the Base Air Staff Officer (BASO). Duties included operational planning and liaison between the units within the Base.

On 5 July he flew in a Lancaster for the first time since leaving No. 617 Squadron. It was a test flight and the crew commented that he handled it well considering how long it had been since he had last flown. On 19 July he joined a Lancaster crew, possibly from No. 630 Squadron, located at East Kirkby, during an attack on a V-1 flying bomb launch site at Criel-sur-Mer in France. He pasted an aiming point photo from the operation in his log book.

On 2 August he was posted to No. 54 Base, RAF Coningsby. RAF Coningsby was a centre for tactical innovation and home of the elite No. 54 Base Flight. Here he was exposed to intelligence that increased his concern that the war would end without him getting back into the action. At this stage he may have had Cochrane's consent for limited operational flying, provided it was non-participatory, short time over target and he could bale out over Allied-controlled territory.

On 15 August he flew in a Lightning as an observer on a daylight raid over Deelen in the Netherlands. He made a similar flight in a Lightning a few days later to Le Havre. On 2 September he flew a Mosquito to Scasta in the Shetlands.

**Final flight**

On 19 September an order came through from Bomber Command for No. 5 Group to prepare for an attack on Bremen. Planes from No. 5 Group would be responsible for all aspects of the operation, including target illumination and marking and control of the raid. Cochrane, the AOC, would be responsible for tactics and route planning. As the day progressed the weather forecast changed, and at 1645 hours an order came through to change to the reserve targets at Rheydt and Muenchen-Gladbach.

At the flight planning conference, it was decided that three areas would be attacked simultaneously; they were designated as red, green and yellow. The red area was Rheydt town centre, where the attack would be fully controlled by a master bomber who would monitor the marking and coordinate the main force bombing. The tactics — dispersed marking — were untried and unrehearsed, and therefore would require expertise from the controller and markers. The announcement that Gibson would be the controller was met with general incredulity. It was assumed it would be regular controller from No. 54 Base Flight or a qualified one from No. 627 Squadron. Some suspicion started to circulate that the proposed complexity may have come from Gibson and his lack of experience in marking techniques.

As Gibson did not have a regular navigator, one had to be found to fly with him. The first choice was ill, so Squadron Leader Jim Warwick was selected. He was the Station Navigation Officer and therefore screened from operations. There was also no serviceable Mosquito available at RAF Coningsby for Gibson to use, so it was decided to use the reserve aircraft of No. 627 Squadron, located at RAF Woodhall Spa. Gibson and Warwick were driven over. When they arrived about 1830 hours, for unknown reasons, Gibson rejected the reserve aircraft KB213 and insisted on using the Mosquito B.XX KB267 instead. The crew who were expecting to fly in KB267 were unhappy with the change. As the two crews were performing different functions the bomb loads had to be swapped. They took off at 1951 hours.

When they arrived at the target, the marking of the red area went badly wrong owing to a series of mishaps. The three markers could not identify the marking point and one aircraft had engine problems. Gibson attempted to mark it himself but his Target Indicators (TIs) did not release. As the illumination from the flares was fading, he called for more flares and warned the red section of the main force not to bomb. He then commanded them to stand by, so they started to turn away from the target. This was potentially dangerous and exposed them to further risk from flak and night fighters. Some started to bomb the green area, either out of confusion or nervousness. He then authorised the remaining aircraft to bomb the green area. The red area was eventually marked, but it was too late to direct any of the main force's aircraft to attack it. The raid concluded at 2158 hours. He remained calm despite the confusion. The time of Gibson's departure from the target is unknown. It is possible that he loitered in a wide, high orbit to assess the outcome and left around 2200 hours. One crew from No. 61 Squadron claimed they heard him say he had a damaged engine.

Gibson's aircraft crashed at Steenbergen in the Netherlands at around 2230 hours. Witnesses heard an aircraft flying low, saw that its cockpit was illuminated and then it crashed. At first, Gibson's failure to return to Woodhall Spa was not considered out of the ordinary, as it was assumed he had landed at Coningsby. Likewise, at Coningsby there was no immediate concern as there was fog and it was assumed he would have landed elsewhere. However, it soon became apparent he had not returned. The rumour spread rapidly around No. 5 Group that he was missing. He was not posted officially as missing until 29 November, although Prime Minister Winston Churchill was informed on 26 September: 'The Air Ministry have told us that Wing Commander Gibson, VC is reported missing from a recent raid in which he flew a Mosquito to Munchen-Gladbach'.

**Funeral**

At Steenbergen, the Germans cordoned off the crash site at the Graaf Hendrikpolder. Human remains were recovered which confirmed there had been one person in the plane and therefore initially it was suspected the other member may have bailed out. However, with the discovery of a third hand, the presence of a second person was confirmed. Jim Warwick was identified from his identity tag. The laundry tag in a sock identified the other person as a "Guy Gibson". The remains were placed in a small specially constructed coffin.

The local deputy mayor, Mr Herbers, wanted to give the men a proper funeral. They hired a horse-drawn hearse from nearby Halsteren. The coffin was draped with the flag of the Netherlands and was laid to rest in the Roman Catholic cemetery. The funeral was attended by the Roman Catholic priest, Father Verhoeven and the Protestant Pastor van den Brink. As they did not know the men's religion, they performed the funeral between them. Father Verhoeven read the psalm, De Profundis and Pastor van den Brink spoke the Lord's Prayer in English. A cross was erected over the grave with Warwick's full rank and name with the name 'Guy Gibson' underneath. When it was later confirmed who 'Guy Gibson' was, a new cross was constructed with Gibson's rank, name and decorations.

The exact cause of Gibson's crash is unknown and is likely to remain so. A number of theories exist, with some more likely than others. Various contributory factors may also have led to the loss of his Mosquito. One theory advanced is that the accident was due to Gibson's lack of experience flying Mosquitos. His log book, which he had last updated on 16 September, detailed 9 hours and 35 minutes flying Mosquitos. It was observed it took him three attempts to land at Scatsta. He had been on one training flight on 31 August to learn how to dive bomb and Mosquito crews knew they had to practise regularly, particularly in pulling out of dives. Also, he had not rehearsed the emergency procedures to exit a Mosquito — as it could take 30 seconds even on the ground.

The same lack of experience flying the Mosquito applied to his navigator, Squadron Leader Warwick, who had never flown in one operationally. That a letter was found with Warwick's address (RAF Coningsby) on it suggests Warwick's inclusion on the flight was a very late decision. He was experienced and would have known not to take anything like an addressed letter with him under normal circumstances.

Harris wrote that Gibson appointed himself as the controller. It is possible he seized this opportunity in Air Commodore Sharpe's absence when the late change in target was announced. There were some instances of Mosquitos breaking up because of their wooden frames. Harris considered this as a possibility, however, it is unlikely. Lack of fuel is the explanation most favoured by members of No. 627 Squadron at the time. In December 1985 the site was excavated and wreckage from the plane recovered. No enemy damage was noticeable. It has therefore been suggested that Gibson and Warwick had failed to switch fuel tanks at the correct time. It has also been suggested there was a fault with the fuel tank selector. Further, it is possible that a lack of familiarity with the Mosquito resulted in neither Gibson nor Warwick being able to find the switches to swap the fuel supply. This would also be a reason to explain why the cockpit was illuminated: they were attempting to locate the switches. In either case, the result would be that the aircraft simply ran out of fuel.

If Gibson left Rheydt at 2200 hours then it is estimated he was about 70 miles short of the expected location if the aircraft had been operating normally. Therefore, it is possible the aircraft was flying underpowered at the time of the crash. This would suggest some sort of damage to the aircraft.

Speculation persists that Gibson's Mosquito may have been shot down by German jet fighter ace Kurt Welter. On the night of the raid, 19 September 1944, Welter was the only German pilot to have claimed a Mosquito shot down that night and Gibson's Mosquito the only Mosquito lost. However, a listing of Luftwaffe claims transcribed from the original microfilms shows that Welter's claim was on the night of 18/19 September, and was north of Wittenberg, in the area southwest of Berlin, more than 500 km from Steenbergen. Welter claimed his Mosquito at 2305 hours near Gütersloh, and recent research indicates that it was actually an intruder Mosquito FB.VI PZ177 of No. 23 Squadron. The crew, Flying Officer K. Eastwood and navigator Flight Lieutenant G.G. Rogers, were both killed.

In October 2011, the Daily Mail featured an article asserting that the cause of Gibson's death may have been a friendly fire incident. Rear gunner Sergeant Bernard McCormack flew on the raid. Before he died in 1992, he left a taped confession with his wife that he believed he had shot Gibson down. He had seen what he thought was a Ju 88 flying near his plane and had fired 600 rounds at it when in the vicinity of Steenbergen. He saw the plane go down. The attack was witnessed by another Lancaster. During the debriefing after the raid he explained what had happened and was asked again about the incident by an Intelligence Officer the following day. He believed the plane shot down belonged to Gibson, as it was not thought that any Ju 88s were present. Reports exist in the National Archive from both crews.

It is possible that Gibson was not where others might have expected him to be. During the briefing for the raid, he was advised to use an exit route that would put him over France. However, he disagreed and insisted he would return by the shortest route at low level.

# The Dortmund–Ems Canal Attack

After the successful Dams Raid 617 Squadron was retained and continued to train with a view toward conducting further special precision bombing missions.

On 8 June 1943 Les was promoted to Temporary Flying Officer in recognition his skill and contribution to the Dams Raid. On 1 August 1943 he was again promoted to Acting Flight Lieutenant.

*Les after being promoted.*

Four months after the Dams Raid eight crews from 617 Squadron were sent out with another new weapon, a 12,000 lb 'thin case' bomb, to attack the Dortmund–Ems Canal.

The Dortmund–Ems Canal was a significant industrial transport route, making it a suitable target. It was believed that if several 12,000 lb bombs were dropped on the aqueduct at low level then it could be breached. These large bombs used had a poor aerodynamic design. To deliver them accurately they had to be dropped from a low level. For this reason, 617

Squadron was assigned the job of destroying the Dortmund—Ems Canal under the codename 'Operation Garlic'. The attack was to undertaken by eight Lancasters supported by six de Havilland Mosquito aircraft from 418 and 605 Squadrons.

The eight Lancaster crews for Operation Garlic reverted back to standard Mk III model aircraft that had not been modified to carry the bouncing bomb. Because these standard aircraft had a mid—upper gun turret an additional air gunner was allocated to each crew, so Les' Dams Raid crew was augmented by Sergeant Leslie Charles Woollard from the RAF. Les Woollard took over from Fred Sutherland in the front gun turret and Fred moved to the mid—upper gun turret.

*Flight Sergeant Leslie Woollard, RAF.*

The attack was launched on the night of 14/15 September however it was recalled while over the North Sea due to fog and mist over the target. Whilst returning, Squadron Leader Maltby's aircraft hit the sea and all the crew were killed. Flight Lieutenant Shannon and his crew circled the wreckage site for two hours waiting for rescue. Later, Maltby's body was the only one recovered.

The attack was re-launched on the night of the 15/16 September. The point of attack was nearby to Munster where the canal divided into two branches. The attacking force split into two groups with three Mosquitos for each four Lancasters. The main purpose of the Mosquitos being to defend the Lancasters from flak.

The weather up to within twenty minutes flying from the target was clear with bright moonlight, but ground fog developed within the vicinity of the target.

Later Rob Kellow recalled the events of the next several minutes: 'We were number six in the order to bomb the target, and whilst waiting our turn were flying on the outskirts of the target at a height of 50 to 100 feet. We had completed two circuits when I felt a violent bump, which appeared to me to come from almost underneath the aircraft. The pilot started at once to climb and when he reached a height of about 500 feet, he sought permission from the leader of the formation—Mick Martin—to jettison his bombs.'

'Two port engines gone. May I have permission to jettison bomb, sir?" he calmly asked Martin. It was the 'sir' that got Martin. Les was following the copybook procedure, asking respectful permission to do the only thing that might get him home. Martin said, 'For God's sake, Les, yes' and as the bomb was not fused Les told Johnson to let it go. Relieved of the weight they started to climb very slowly...

Les continued to climb, but as the two port engines showed signs of catching fire they were feathered. After reaching a height of approximately 1500 feet the starboard inner engine showed signs of catching fire, and was also feathered. The starboard rudder also appeared to be damaged. The controls were getting worse all the time until, though he had full opposite rudder and aileron on, Les could not stop the aircraft turning to port.

The Lancaster then started to lose height gradually and it became clear to Les that he could never fly her home. At about 1200 feet, Les ordered his crew to bale out and held the plane steady while they did. Above Amelo, Holland the crew members all then made escape out of the damaged aircraft.

The scene inside the aircraft just before the crew began baling out was further described by wireless operator Bob Kellow in his memoirs:

'We had crossed the Dutch/German border and were about half way to the Dutch coast. We all knew that at this height and with only one motor working properly our chances of getting back to England were slim.'

Les had asked our rear gunner, 'Obie' O'Brien, to go to the front gun turret...

'OK I'm in the turret, Les. What do you want me to do?'

'Good, now reach along below my feet Obie and see if you can find a loose, broken cable,' said Les. 'It belongs to the starboard rudder. When you find it, pull on it for all you're worth.'

In a few minutes Obie announced that he'd found the cable and was pulling it.

The plane began to swing slowly to the right. It was only then that I realized that we'd been steadily swinging to the left for the past few minutes.

'I'll have to stop the starboard inner, Les,' said Ray, our flight engineer.

'Try to hold it a bit longer, Ray,' Les replied.

Obie meanwhile warned that his arm was breaking from pulling on the cable and he'd need a break.

'OK Obie, but pull on it again as soon as you can,' said Les.

It was clear Les was putting on a superhuman effort to keep our crippled plane on some sort of course, but I knew we couldn't go on much longer. The plane was down to 1,000 feet and the glide angle was steadily increasing.

'Send out that we're baling out, Bob,' Les said to me.

I unhooked my Morse key and began tapping out the message. Before leaving the aircraft, I did hear messages received from two of the aircraft wishing us good luck.

The crew prepared themselves, and one by one they left the aircraft. Kellow moved forward to the cockpit.

'I stood by him as he firmly held the wheel and tried to keep the aircraft on a steady course, making it easier for each man to jump out. Like a sea captain, he wanted to be sure everyone was safely off before he abandoned ship. His parachute was clipped onto his harness and he looked searchingly at me, probably wondering why I hadn't jumped already.'

Using signs, I asked if he was OK. He nodded his answer and a wry smile puckered his mouth.

With a last smile, I gave him the thumbs-up sign, checked my parachute and took my position at the edge of the escape hatch. Then I bent forward with my head down and tumbled out into the darkness.'

Les stayed at the controls of the stricken Lancaster and changed course to avoid crashing into the built-up area of the village of Den Ham. He attempted a forced landing in a field near Den Ham at 0346 hours. He nearly succeeded, but the aircraft hit a bank running across the field and burst into flames. An eye witness, Henk Kremer, who was an eleven-year-old boy in 1943, wrote the following account in 2018:

> 'At about 3.30 am my father woke me telling me that a burning aircraft was flying towards the village. When I looked outside my bedroom window, I saw on the east side of our village a low flying burning aircraft, it was flying towards the village. I remember thinking: this is not going to end well. At that moment, I saw the machine make a slight turn to the right changing his flightpath to a northerly direction. Straight away the aircraft made another course change by turning sharply to the left... I saw the fire at the front of the aircraft had become fiercer. I saw the aircraft quickly lose height and that the propellers were ablaze. Then the aircraft crashed, and then there was only an intense fire was visible. This happened about 1200 metres from our house.'

*The wreckage of AJ–N on the morning of 16 September 1943.*

Another witness, Bertha Bakker, then a teenager, and whose family owned the land next to the crash site, takes up the story:

> 'My father went straight to the crash site. My sister and I just followed my father... The heat was terrifying and very intense. It was terrible to see. I was maybe 100, 200 metres from the crash... My father was there quicker than the Germans. My father saw Knight in the cockpit and he was crouched over. He was leaning forward. He was not sure but it looked as if he was crouched over in the seat trying to cut himself out of the safety harness. My father saw him burning. It was horrible, just horrible.'

It's clear from several witnesses that Les deliberately steered away from the centre of the village in an effort to avoid killing any civilians. Another young boy who visited the crash site, Lucas Kamphuis, has said that he was 'an exceptional person to have the clearness of mind to do what he did.'

The next morning, the Dutch police and the German military cordoned off the area. Les' body was removed from the wreckage and a local schoolmaster Derk Jan Snel took a risk by taking a photograph as this was done. He later took another of the hearse taking Knight's body to the cemetery.

The body was taken by horse—drawn hearse to the old cemetery in Den Ham by local funeral director Gerrit Meijer, who led the cortège on foot. Dozens of local people gathered in the streets and at the entrance to the cemetery to pay their respects, although the Germans prevented them from entering the cemetery. One young boy, Henk Steen, however took a chance and climbed through a hole in the hedge:

> 'I stayed very close to the hedge. The German soldiers saw me but did not send me away. I stood maximum ten metres from where everything was taking place. I saw six soldiers march into the graveyard with a German officer and [preacher] Dominee Meuleman. Three soldiers stood on either side of the coffin, Meuleman said some prayers and the officer spoke of course in German. I heard him clearly but did not understand much of what he said. I was told later by someone who could understand German that the officer said that he saw Les

# The Dortmund—Ems Canal Attack

*The carriage taking Les to the Den Ham Cemetery. Note the significant number of local onlookers.*

*Les Knight's gravesite in 1943. The details on the cross have probably been gleaned from his identity disc.*

Knight as a 'brother-in-arms and not as an enemy'. The officer then ordered the soldiers to shoot a salvo as military salute. This of course was a very honourable thing to say. For years I have thought about what the officer said and conclude: I believe that officer was a good man. There was no way that a German officer would stay anything as noble as that.'

Les' grave was first marked with a simple wooden cross, which was replaced after the war with a Commonwealth War Graves Commission gravestone.

Of the eight Lancasters that departed that evening for the attack only four returned. The Dortmund-Ems Canal was not significantly damaged by the attack, and 617 Squadron lost five of the eight aircraft involved in Operation Garlic. All the Mosquitoes, however, returned safely.

The RAF casualty notification system kicked into effect and within hours Les' family were informed that he was missing and then later that he was presumed dead.

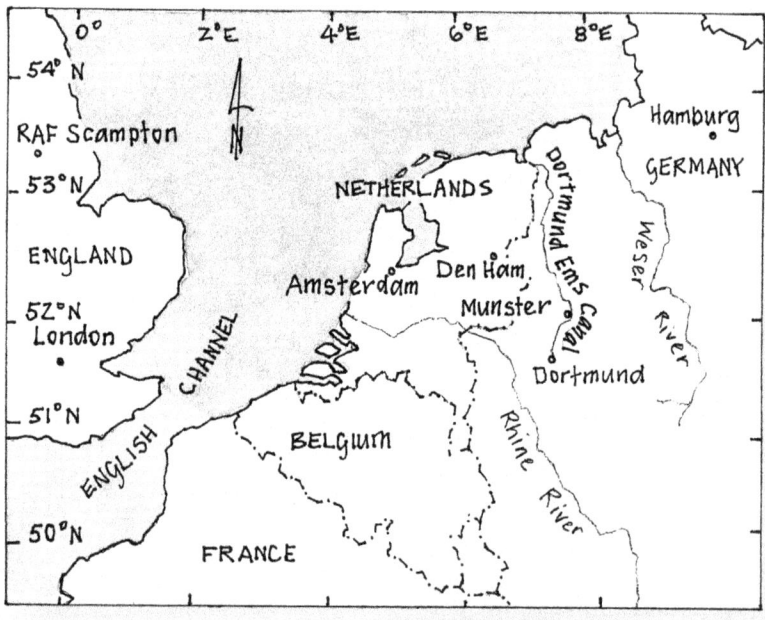

# LES' LEGACY

All seven of Les' crew mates were able to open their parachutes and given their low altitude made hard landings. They had all been separated and they were now alone in the dark in enemy controlled territory. And because they had been circling in a burning aircraft, they knew that the Germans would have seen them parachuting out of the aircraft.

Ed Johnson baled out yelling 'Cheerio boys. Best of luck. See you in London.' He recalled later 'The farewells were a little hasty but lacked nothing in sincerity for that.' Johnson landed safely and successfully evaded capture and reached the safety of Spain with the help of a friendly Dutch farmer and policeman, and various members of the resistance in Holland, Belgium and France. He returned to the UK via Gibraltar. Johnson served out the rest of the war in various ground postings and left the RAF in 1947. He returned to Blackpool, and joined a company selling fireplaces, where he worked until his retirement. Johnson died at age 90 at Blackpool, UK on 1 October 2002.

Harold Hobday was able to evade capture. Within a few hours he had made contact with Dutch resistance supporters. He was taken to a woodland shack near Baarn and reunited with his colleague Fred Sutherland. The pair were then fed into the escape network and smuggled the whole way through France to the Pyrenees, then onward through Spain to Gibraltar and then returned to the UK. After the war, Hobday returned to Lloyds' and eventually rose to become head of the aviation department. He died at age 88 at Fakenham, Norfolk on 24 February 2000.

Ray Grayston landed safely but was captured almost immediately. He was sent to Stalag Luft III until January 1945 when the POWs were forced to march westwards. He reached Stalag IVA at Luckenwalde where, after three months he was liberated and flown back to England. Grayston left the RAF after the war and joined the aviation firm Hawker Siddeley. He worked as a quality inspector and retired in 1984. He died at age 91 at Woodhall Spa, UK on 15 April 2010.

Rob Kellow landed safely and was able to evade capture. He also entered the escape network and made the six-week journey through France to Spain a day or so ahead of Hobday and Sutherland. Kellow arrived back in England in December 1943. Like all those who had made the perilous return trip he was not allowed to fly over enemy territory again lest he be recaptured and forced to compromise the escape network. Kellow returned to Australia in May 1944 and served in the RAAF's 37 Squadron for the remainder of the war, mainly in Australia but also including a deployment to New Guinea.

After the war Kellow returned to Newcastle after being discharged from the RAAF with a glowing report from his commanding officer who described him as showing 'great possibilities for good leadership' and 'one of the most well liked and well known' and an 'invaluable' member of the Squadron. Kellow returned to his job as a shop assistant and in 1946 married Doreen Smith, a Canadian who he had met while training there in 1941. By 1952 they had two children and the whole family moved to Winnipeg, Canada, where Kellow worked for the Manitoba Power Commission. He travelled back to the UK for a number of 617 Squadron anniversary events, and paid his respects at Les Knight's grave in Holland. Kellow died at age 71 at

Winnipeg on 12 July 1988. After his passing a memoir of his evasion and escape was published titled *Paths to Freedom*.

Fred Sutherland landed safely and was hidden by a Dutch farmer. He was then put in touch with the Dutch underground network and met up with Sidney Hobday. The two were smuggled all the way through Belgium and France to Spain. Not long after getting back to the UK Sutherland was sent home to Canada. Greeted in Edmonton by his girlfriend, Margaret Baker, he immediately proposed to her.

After the war Sutherland stayed on in the RCAF and was commissioned. He then studied forestry and got a job with the forestry service. In 1964 he became the forestry superintendent in Rocky Mountain House in his home province of Alberta. Sutherland died in late 2018 at the age of 95 in Rocky Mountain House near Red Deer, Alberta. Fred was the second last Dambuster alive.

*Fred Sutherland in 2017.*

Henry O'Brien landed safely but was captured and spent the rest of the war as a POW. O'Brien returned to Canada after the war. He died at age 62 on 12 September 1985 at Edmonton, Alberta.

Les Woollard landed safely and was hidden by a Dutch farmer and his family for ten days. With the help of the local resistance members he was moved onto the nearby town of Meppel where he was hidden in an old potato warehouse for two weeks. He was then put on a train with a sign hanging around his neck stating: 'Deaf and Dumb' and later hidden in a secure house in Paris which was situated next to a Gestapo HQ. From Paris he was put on a fisher's boat at Calais. He got back to the UK in time for the birth of his daughter. Woollard died at age 57 in September 1978 at Lewes Sussex.

Were it not for Les Knight's courage to allow them to all parachute out of the stricken Lancaster the whole crew might very well have died that night. But together, the seven members of Les' crew lived for another 380 years. And each of them was able to marry and have children. There is no doubt that Les' crew mates all owed their lives to their young pilot, something that they never forgot. In all, a magnificent legacy for a brave Australian pilot.

As the five crew members retuned to England they were required to report on their experiences. As the accounts came together Les' heroic actions became very clear. The RAF realised that Les' actions needed to be recognised and so on 1 January 1945 he was awarded the Mention in Dispatches 'in recognition for distinguished service'.

# Remembering Les Knight

A few months after Les was buried in Den Ham on 10 December 1943 a Mosquito of 487 Squadron, RNZAF was shot down and crashed south of Den Ham. Flight Sergeant Thomas Mair, RAF (aged 24) and Warrant Officer Kenneth Blow, RNZAF (aged 22) were both killed. They were buried alongside Les in the Den Ham Cemetery.

After the war the Commonwealth War Graves Commission was able to register the three places of burial. A white cross was placed in front of the wooden cross showing Les' full service details. In time the crosses were replaced by the standard Commonwealth War Graves Commission headstone.

In 1954 Les' mother was able to visit Les' grave. By this time the story of Les' heroism was becoming better known. During her visit it is understood that she was hosted by several former members of the Dutch resistance who helped Les' crew mates to avoid capture and eventually escape back to England.

In the late 1980s Les' former crew mate and fellow Australian Rob Kellow visited Den Ham to pay his respects.

Vera Newton (who was introduced in Chapter 1) visited Les' grave in 2013. Vera remembered the day after the Dam Busters raid when Les turned up unannounced. 'He still had on his flying boots and jacket and looked

*Rob Kellow visits Les' grave in the 1980s.*

shattered,' she said. 'My mother packed him off to bed and no questions were asked.' As news of the British success on the Eder Dam filtered through, her father broached the subject with the exhausted young bomber pilot. 'When Les got up the following day, my Dad asked: 'Were you on that, son? and all Les said was: 'Yes. That was it.'

*Vera Newton visits Les' grave.*

Over the weekend of 14–16 September 2018 Den Ham locals organised a weekend commemoration of Les Knight, bringing together members of his family, the families of his crew. Also present were family members of the local underground resistance movement which helped several of the crew evade capture and return to England.

# Remembering Les Knight    135

Les Knight's grave with an additional
Commonwealth War Graves Commission
cross circa 1945.

Les Knight's gravesite with the
Commonwealth War Graves Commission
headstone.

## DEN HAM GENERAL CEMETERY
(Index No. NL. 382)

Den Ham is a large village 16 kilometres (10 miles) north-west of Almelo, on the road to Meppel and the north. The most convenient railway station is Almelo with a connecting bus service to the village. Accommodation is available in Almelo, and there is good accommodation in Den Ham also. The cemetery is about half a mile south-east of the village, on the north side of the road to Almelo. In the south-western part are the graves of 3 airmen, two from the United Kingdom and one from Australia.

BLOW, Wt. Offr. (Nav.) KENNETH LESLIE OWEN, 751684, D.F.C. R.A.F. (V.R.). 487 (R.N.Z.A.F.) Sqdn. 10th December, 1943. Age 22. Son of Edward and Annie Elizabeth Blow, of Luton, Bedfordshire. Grave 763.

KNIGHT, Flight Lieut. LESLIE GORDON, 401449, D.S.O. Mentioned in Despatches. R.A.A.F. 16th September, 1943. Age 22. Son of William Henry Harold and Nellie Marsom Knight, of Camberwell, Victoria, Australia. Grave 765.

MAIR, Flt. Sgt. (Pilot) THOMAS, 656406. R.A.F. 487 (R.N.Z.A.F.) Sqdn. 10th December, 1943. Age 24. Son of Marion Mair, of Greenock, Renfrewshire. Grave 764.

*The three Commonwealth War Graves Commission graves in the Den Ham General Cemetery.*

*Les' mother Nellie visits his grave circa 1954. It is believed she is accompanied by former members of the Dutch resistance.*

*In 2002 three local men — Piet Meijer, Herman Stegeman and Lucas Kamphuis — established a granite memorial at the crash site. The memorial stone is now recognised as a national monument.*

In Australia, Les's family planted an Irish Yew Tree in their back yard and attached a memorial tablet to it. In September 1952 at the Camberwell Methodist Church a memorial brass plaque was unveiled and remains there to this day.

*Locals commemorating Les Knight's death in September 2018.*

# Conclusion

War is an unfortunate feature of the human condition. For millennia it has been responsible for suffering and loss but it also remains as a tool of statecraft and politics and its eradication has proven elusive.

And despite the misery war invariably generates it is also capable of bringing out both the best and the worst in individuals.

Many would consider that training and equipping young men to drop bombs from planes high above the ground to be obscene and an evil act by any government—particularly if the distinction between bombing 'legitimate' military targets and innocent civilian targets is knowingly impossible to achieve.

Similarly, others might justify the means as a necessary way to achieve the ends—to defeat fascism. There is no right or wrong answer and it is a topic that must be grappled with each time humanity embarks on a war.

What is more well understood however is that humans are gregarious and like to work towards goals in teams. They gain satisfaction from knowing others and together achieving what they would not be able to alone. This is what most soldiers find as a positive memory from their service in the Army, Navy or Air Force—on either side of the conflict. And paradoxically, the extraordinary challenge of war is known to bring out the best qualities in people—courage, self-lessness and compassion.

The counterpoint is that fighting in a war is a dangerous business and killing others is part of the calculus of the struggle. Killing other people does not sit comfortably with human nature. Some are more inured to it than others but many will carry the memory of killing another person like a stone for the remainder of their lives.

Even more damagingly, having or seeing your comrades killed or maimed is another emotional burden that time can fade only slowly.

So would Les Knight be accused of murdering innocent people—some would say he could. Could he be simply regarded as doing what his country asked him to do in order to win the war against Nazi Germany—indeed, like many thousands of other young men and women. Did he care for the other men in his crew and do his utmost to ensure their collective survival by doing his job to the best of his ability—it certainly seems so. Did he display courage in the face of extreme adversity—very much so. He lived in challenging times and he experienced more life by the age of 22 than many others. We now need to try to understand those times, the challenges and the achievements of that generation.

Les' parents requested a short epitaph be carved into his headstone in that small village in Holland—'Let us be worthy'. I like to think they hoped that we would strive to remember and emulate Les' courage, selflessness and sacrifice.

Marcus Fielding was born and raised in Melbourne. He joined the Australian Regular Army in 1983 and graduated from the Royal Military College Duntroon as a Lieutenant in 1986.

In the following decades of military service Marcus held a broad range of senior appointments in Army, defence and interagency organisations in a number of locations throughout Australia and overseas.

Marcus has participated in four operational deployments. In 1992 he directed operations to clear land mines in Afghanistan. In 1995 he coordinated infrastructure construction projects in Haiti.

In 1999 and 2000 Marcus directed security operations and coordinated the repatriation of displaced persons as part of the Australian-led international force in East Timor. For his work in East Timor, he was awarded a Commendation for Distinguished Service.

In 2008 and 2009 Marcus spent nine months in Baghdad as an 'action officer' in the Headquarters Multi-National Force–Iraq. In 2011 he published a book about his experiences in Iraq titled *Red Zone Baghdad*. Colonel Fielding transferred from full-time to part-time service with the Australian Army in 2011.

Marcus is the President of the Camberwell City Sub-Branch of the Returned and Services League of Australia and of President of Military History and Heritage Victoria—an inclusive forum for individuals and groups who are passionate about military history and heritage.

www.ingramcontent.com/pod-product-compliance
Lightning Source LLC
Chambersburg PA
CBHW071411160426
42813CB00085B/956